DEATH BECOMES THEM

ALSO BY ALIX STRAUSS

Fiction
The Joy of Funerals

Nonfiction
Have I Got a Guy for You

DEATH BECOMES THEM

Unearthing the Suicides of the Brilliant,
the Famous, and the Notorious

Alix Strauss

HARPER

NEW YORK · LONDON · TORONTO · SYDNEY

HARPER

FIRST EDITION

Designed by Justin Dodd

Library of Congress Cataloging-in-Publication Data is available upon request.

ISBN 978-0-06-172856-3

09 10 11 12 13   OV/RRD   10 9 8 7 6 5 4 3 2 1

For my parents,
who, thankfully, are very much among the living

Razors pain you;
Rivers are damp;
Acids stain you;
And drugs cause cramp.
Guns aren't lawful;
Nooses give;
Gas smells awful;
You might as well live.

Dorothy Parker, "Resume," 1926

Though known for her biting wit, her edgy personality, her love of drinking, and her membership in the famous Algonquin Round Table, Dorothy Parker tried to kill herself four times. Severely disappointed and dissatisfied with the freelance magazine life, and having both money problems and an array of failed relationships, she made her first attempt after an abortion in 1923 when she cut her wrists. She also overdosed on the sedative Veronal, consumed a bottle of shoe polish, and took sleeping powder. On June 7, 1967, she was found dead in her hotel room at the Hotel Volney from a heart attack. She was seventy-three.

Contents

Acknowledgments

An enormous thanks to:

The always amazing, never tiresome team at William Morris who champion my projects and efforts: Lauren Heller Whitney, Alicia Gordon, Anna DeRoy, Bethany Anne Dick, Caroline Donofrio, and especially Andy McNicol—who read, responded to, repositioned, and redesigned my literary endeavors.

The wonderfully smart peeps at HarperCollins: Sharyn Rosenblum, Jamie Brinkhouse, Blair Bryant Nichols, Carrie Kania, Vanessa Schneider, and my brilliant editors and friends Jennifer Schulkind and Mauro DiPreta—both of whom not only appreciated my dark side and quest for things not so pretty, but also made this project something I could be proud of.

My researchers, who each helped make this come together in their quest for odd facts and their eye for detail, and who I'm hoping, if they're ever contestants on a game show—should they ever want to be contestants on a game show—will be able to use the information they gained on this project to win oodles of money: Geoffrey

Kellogg, Leigh Malach, Nicole Robson, Anastasia Dyakovskaya, and Natalie Zutter.

The specialists interviewed: Jonathan Alpert, Leo Braudy, Joshua Gamson, Dr. George E. Murphy, Mike Murphy, Dr. David Lester, Rebecca Roy, Daniela E. Schreier, Dr. Edwin S. Shneidman, Doug Thorburn, and Dr. Paul Zarkowski. Your words of wisdom were invaluable.

The Grave Spotters: Ben Ellis, Kevin Dawson Jones, Danny Garside, and Michel Enkiri, for their fabulous photos.

The PR team at LaForce & Stevens: Leslie Stevens, James LaForce, and Meryl Weinsaft Cooper, who always astound me with their generosity and knack for all things clever.

My writer friends/advisors: Charles Salzberg, for our Sunday night dinners; Marci Alboher for her positive outlook; Lisa Rosenstein for her love, support, and endless phone calls; Jami Beere for her constant presence and for working vigorously as my PR agent; and Rebecca Mattila and Andy Christie for their wonderful web work and, more importantly, their friendship.

Lastly, a respectful nod to the souls we highlight here (and the family, friends, and fans they left behind). You are all missed.

Introduction
Unearthing Greatness

"In the winter everyone is matted down in thick wool coats as we stand in a huddled mass of sniffles and tears. In the summer the warm air rings full of sorrow as mourners sigh in sadness, cotton jackets and black dresses blowing in the breeze. To me, it doesn't matter what season a funeral takes place, I enjoy them just the same."

Thus began an article I wrote for the "Lives" section of the *New York Times* almost a decade ago. I'm as fascinated by funerals today as I was then. I blame my odd attraction to death and memorials on my being an only child—actually, the *only* only child in my family. For as many generations as I can trace back, everyone has had several children—except for my parents, who decided to have just me. Growing up, there were no holiday dinners spent bonding over burnt turkey and overcooked stuffing, no long-distance, late-night phone calls, no group vacations with family members. And so funerals became my only chance to bond with my relatives, many of whom I'd never met. Rather than a solemn event, I regarded them as reunions.

Paying respect at a relative's home became like trying to find a secret stash of candy. I couldn't help but inspect each room, search through the cabinets and dresser drawers, peer at scrapbooks and photo albums filled with old Kodak memories. I snooped in the hopes of finding answers to who they were; I searched for something that would connect me to them. To make me understand. It is the "Where do I belong?" and "Where do I come from?" that was missing from my life. This longing for a connection to someone or something is a feeling I have never been able to let go of.

Funerals are an often misunderstood societal phenomenon. The topic, once considered forbidden or taboo, has now become trendy. As a culture, we are obsessed with death. As a population, we connect with one another by sharing the same experiences. Misery loves company, and company is what we crave no matter what our nationality or religious beliefs. It's why we can bond instantly with strangers as we stand swaying during an all-night vigil, lit candles illuminating our faces and the faces of our new acquaintances. Kurt Cobain's public vigil was held at Seattle Center's park and drew approximately seven thousand mourners. Prerecorded messages by Courtney Love and Nirvana's bassist, Krist Novoselic, were played. Love read portions of her husband's suicide note to the crowd and, at the end of the ceremony, gave away some of his clothing to those who remained.

Years later, these moments will become our earned badges of "mourning memory," which we will share at bars, cocktail parties, and random events while reminiscing about the departed, sharing where we were at that moment Ernest Hemingway shot himself; when Diane Arbus filled a tub with warm water, swallowed a handful of barbiturates, and then slit her wrists; when Spalding Gray went missing, and when they fished his body from the East River two months later. It is the "I was part of that, I was there" that we yearn for.

We are also addicted to the drama. We crave their stories the same way they craved their pills, liquor, coke, and heroin. We want to understand the sadness they felt and the depression they couldn't live with. Our insatiable preoccupation with celebrities has been heightened thanks to our morbid fascination with how they died. Add a suicide, and our quest for more is as strong as our need for air. As tempting as a letter marked "Do Not Open."

And there is loss.

We have indeed lost something great and historical, important and special, in each, be it a rock star or writer, poet or politician, activist or artist, singer or starlet.

Why *do* we love these tortured souls? What is it about their suicides that is so intriguing? Did they achieve celebrity status for their body of work or did they become even more famous, reaching a higher iconic status, only after killing themselves? Vincent van Gogh sold just one painting while alive. After he killed himself, he and his art became legendary.

Long ago, self-murder was often viewed as a noble way to defy persecution while receiving notoriety for standing up for one's principles. Think Socrates, Cato, and Seneca, each of whom chose suicide as a way to free himself. But times have changed, and now suicide can be a fast track to fame. Instant publicity forever encapsulated by a signal event.

Along with being a captivating overview of suicide, this book will delve into twenty memorable ones. Each death is as diverse as the person who killed himself. Some are tragic: Dorothy Dandridge was found naked on her bathroom floor, a handful of antidepressants swimming in her system. Others are bizarre: Hunter S. Thompson shot himself while on the phone with his wife. But all are memorable.

This book is also a tribute, meant to recognize their notable achievements and acknowledge how difficult it must have been to produce and create under sometimes dire circumstances. Their stories are to be handled like eggs, carefully and with kindness. It's a front-row seat on the lonely, personal nightmares experienced by these legendary luminaries. It concentrates on their final days and the incidents that led up to the moment when they took their last breath, followed by the moment when they no longer could.

Selecting which icons to acknowledge was not easy. Sadly, there were way too many significant suicides to choose from. From the most loved to the most feared, countless personalities have affected our culture and inspired and enthralled us. The people highlighted here were considered essential for a variety of reasons: Sigmund Freud, for being one of the first major documented assisted suicides; Peg Entwistle, for her choice of location; Adolf Hitler, for his enormous impact on history; David Strickland, for ending his life sadly, lonely, and strangely, at the height of his success as an actor.

In most cases, those presented here experienced clear, defining moments or left traceable, tangible pieces of evidence: a suicide note or the giving away of prized possessions, which helped prove that their deaths were intended rather than unintentional. Virginia Woolf put stones in her pockets and then walked into a river, her last words in a note to her husband placed on a mantel in her home. Sylvia Plath left a note and a manuscript; after preparing food for her children and setting it by their beds, she sealed their room, and then her kitchen, turned on the gas oven, and stuck her head in it.

Celebrities who died from an accidental overdose rather than a purposeful decision to end their lives—Hank Williams, Judy Garland, Elvis Presley, John Belushi, Edie Sedgwick, Frida Kahlo, Jimi Hendrix, and, most recently, Heath Ledger—didn't fall into the suicide category. Some had death wishes—James Dean. And then there were those whose departures remain mysteries: Marilyn Monroe and George "Superman" Reeves. Murder or suicide? No one knows for sure.

When people heard the title of this book, they were fascinated, eager to learn more, asking for facts I'd uncovered or why this person was chosen as opposed to that one. The truth is, people want to know the gritty, dirty details. They want to unearth the not-so-pretty picture while taking a close look at the bag of bones left behind. It's human nature to ask what happened upon hearing someone has killed him or herself. "How did they do it?" and "Why did they do it?" are often the next questions that slip from their lips when they learned suicide was the culprit.

While writing this introduction, I realized that examining these great people was no different from snooping as a child through the rooms of relatives who had died. Each time I hoped to gain a deeper understanding of my family. I do the same now as I examine these nonconformists. They are the rooms I've yet to inspect and the secret compartments that still need opening. Through analyzing their suicide notes, the clues they left, the people they interacted with, and the bizarre and strange behavior they displayed, I hope to gain understanding. It's the puzzle not yet finished, the questions that still need answers. And of course, there needs to be a final goodbye.

DEATH
BECOMES
THEM

One

Grave Intentions

Suicide has many names: solitaire, intentional self-killing, self-inflicted fatal wounds, and dirt nap. Cops refer to self-murder as "doing a dutchie" or "taking the night train." In Britain, suicide is called "topping oneself" or "soap suds." It doesn't matter which term you use, the outcome is the same. Many say that suicide is a permanent solution to a temporary problem. Each year in the United States more than thirty-two thousand people succeed at it. Eighty-six Americans do it every day—which works out to one death every sixteen to eighteen minutes. Throughout the world, about two thousand people kill themselves daily. That number more than doubles for unsuccessful suicide attempts.

There is something devastatingly sad about a life ending too soon. It's the here, and then the not. The clock you can't turn back. For survivors, those left behind, it's the heartache. The life you couldn't save. The constant inner nagging that you didn't do enough, paired with the unfortunate realization that you did all you could.

More so than ever, celebrity suicide has become a politically hot subject, and we have become almost morbidly curious. "I would love to think that the culture's fascination is because Plath is a great and major poet, which she is," poet and critic Al Alvarez once said of his close friend. "But it wouldn't be true. It is because people are wildly interested in scandal and gossip." Why are we so fixated on these famous suicides, on these brilliant, tortured souls? No detail is too small for our consumption. No information too ghastly for us to hear.

"We are interested in the particulars because we are searching for understanding and for something to connect to," says Rebecca Roy, an entertainment industry specialist and psychotherapist based in Los Angeles. "It's projection. All of these people embody something emotionally that resonates for us. We project our own hopes, dreams, and fears onto famous figures, and we then get to watch them played out in front of us."

Knowing the facts surrounding the death of poet Anne Sexton, who draped herself in her mother's fur before climbing into her car to inhale a garage full of carbon monoxide, helps us understand her pain while feeling closer to her.

"The fur was obviously important, and clearly there was a conflict with her mother," says Roy. "Who hasn't felt that? Who wasn't able to make a relationship work? When we see a piece of ourselves in the story, it intrigues us because we, too, are struggling with those issues. We respond to the story as it propels our transference of our own feelings onto that other person. So when a celebrity that we've been invested in dies, it feels like a huge tragedy. Like a piece of us has died as well."

Through these celebrities' artistic endeavors, their writing, their songs, their appearances on television or in films, or because of the impact they made on history, we feel joined to them somehow. Sometimes we feel closer to these strangers than we do our own neighbors, friends, and family. We let these larger-than-life personalities into our homes, our lives, and our hearts. We think we've earned the right to mourn because we feel incredibly and indelibly linked to them. The more we see, the more we witness, the closer and more united to them we feel. Thanks to twenty-four-hour TV coverage, gossip magazines, newspapers, websites, and reality shows, we've become an instant-gratification media generation, constantly exposed to the celebrities we idolize—all of which has fueled this sense of false intimacy.

Many of us have followed celebrities' careers. We have watched them fall in love, get married, have children, and become divorced. We have seen them struggle publicly as we struggle privately. We have been assigned their books in school, bought their CDs, memorized their lyrics, even quoted them in our yearbooks. Since we have a front-row seat into their lives, we are there to champion their triumphs, and share their hurt and disappointment when they are defeated. We witness their awards speeches, we are present at their concerts and theater performances, we have seen their masterpieces hanging in museums, and we've attended their readings at bookstores. We have gathered at their political rallies, listened to their lectures, and watched the way they've changed the world.

Aside from suffering from projection, there's the fairy dust phenomenon to contend with. "We assume if we're near these celebrities and famous figures, their magic or good fortune will rub off on us," Roy explains, adding that we think that by reading their books, watching their interviews, and learning as much as possible about them, their secrets for success will be revealed to us.

"Given our obsession with celebrities and interest in things that are taboo, and the fact that they died and that it was a suicide, gives their death some significance and thus, a heightened interest," says Dr. David Lester, Ph.D., a suicidologist who has studied self-killing for over forty years. "Most people have suicidal and self-destructive impulses and spend a lifetime repressing them. These celebrities acted on them. That, paired with their fame and our feeling of connection, heightens the interest we already have."

And we need them—to show us that the impossible is not, to utter what we are unable to verbalize. They are underdogs overcoming adversity as we watch them do it. We want to save them, and yet be saved by them. So we allow them their bad behavior, grant latitude to their extreme mood swings.

We also love their brilliance and their genius. The contributions they made to history are fingerprints carved with a sharp knife, its indentation a valley of inerasable crevices. And so, the loss seems that much sadder, its impact that much greater. We fall prey to their good looks and their artistic talents. We're dazzled by their glamorous lives and wooed by their entrance into an exclusive club we long to be members of.

"When they kill themselves, our shock factor is heightened. We can't fathom how someone with wealth, beauty, and fame could be so miserable. We think someone with all those external achievements has to be happy," Roy says. "When we see someone who we think has everything going for them, and they kill themselves, it confuses us. It throws our values into question." And disrupts our belief that fame cures all ills. What people really desire, we think, is validation. We want to be wanted and loved. When someone has gotten that validation, and still commits suicide, we are puzzled. We need to know more.

Suicide and darkness have long plagued the ultra-creative. Their self-destructive vices and passion for excess follow them like a trail of empty bottles, and often beg the chicken-or-egg question. Is it their sadness that makes them so brilliantly creative, or does their brilliance and ability to create induce their sadness?

Dating back to Plato—who often spoke about creative individuals and how they were susceptible to melancholy—we have separated clinical insanity from creative insanity. Seneca is often quoted as saying, "There never has been great talent without a touch of madness." Even nineteenth-century essayist Charles Lamb noted the too-close-for-comfort connection in *The Sanity of True Genius*: "So far from the position holding true, that great wit (or genius, in our modern way of speaking) has a necessary alliance with insanity; the greatest wits, on the contrary, will ever be found in the sanest writers."

There's no disagreeing that death obsessions, pain, and deep complexity are themes of artistic geniuses' work, that they define their material. When famed abstract artist Mark Rothko gave a lecture at the Pratt Institute on the ingredients and recipe for making a work of art, he stated, "There must be a clear preoccupation with death." He continued by adding in some sensuality, mixing in tension and irony, wit and play for the human element, and a touch of the ephemeral, and chance. Stir and let sit, then end with hope. "Ten percent to make the tragic concept more endurable," he insisted.

The failure of his band Attila led Billy Joel, a then depressed alcoholic, to attempt suicide in late 1970 by drinking furniture polish. "It looked tastier than bleach," he shared in Hank Bordowitz's biography of the singer-songwriter, *Billy Joel: The Life and Times of an Angry Young Man*. The suicide note he left later became the lyrics to his song "Tomorrow Is Today."

And there isn't an art historian who will deny that van Gogh did some of his most impressive, most important work—like many others in this book—while in the throes of a deep depression. *Starry Night* was created while van Gogh was in a mental institution. "The more I am spent, ill, a broken pitcher, so much more am I an artist, a creative artist," he once admitted, adding that he put his heart and soul into his work, "and have lost my mind in the process."

Much of their selves ends up in their opuses. They vomit up their feelings, hoping it will empty them out. And yet, each morning or evening, after a drink or two or three, after the pills have stopped working and the drugs have worn off and they've returned from Oz, they are still filled with pain, stuck with their inner devils and demons. After choking on their sadness and drowning in genius, suicide seems like a suitable solution, instant relief from a lifetime of agony. More than the body of work an icon creates, what will forever define him becomes his suicidal act. The intriguing stories around his death make him a shadowy figure who lurks in the forefront. A ghost who hovers.

Our attraction to famous figures is a relatively new phenomenon, occurring over the past two hundred years. "It started with the American Revolution," cites Leo Braudy, a cultural historian and author of *The Frenzy of Renown: Fame and Its History*. "We're the first country that existed on purpose. We decided what our flag would look like and what's our symbolic code. Everything was done from scratch and so we developed heroes."

Ben Franklin, George Washington, Thomas Jefferson—our first American Idols, so to speak—received VIP treatment while being indoctrinated with star status. Braudy says that this phenomenon is "in a way similar to the treatment that Louis XIV gave to himself, but we did it for democratic reasons, and was the unprecedented fame of how celebrity evolved."

In the 1920s and '30s, film studios tried to keep any gossip or negativity concerning its contract actors out of the papers. Today that information is an important part of celebrity journalism.

"People have been interested in powerful individuals for centuries," adds Joshua Gamson, author of *Claims to Fame: Celebrity in Contemporary America*. "The interest hasn't changed, but the supply of people has. Today, when anyone can be a star, there's less distance and we can identify with them more easily, so our attachment is increased. We think, 'They're like me.' It's like a friend or a peer dying rather than a God."

Gamson also differentiates between what we call a celebrity, and heroic figures who did something extraordinary and contributed to our nation. "Those people are the more traditional kind of fame created from merit or spectacular contribution, whether positive or negative," he says. "Their fame is tied to their achievement, which is different from today's media-generated celebrity or where people are famous for being themselves. In this case, the loss for us tends to be more fascinating and adorning." Rather than merely feeling we knew them—as we do with celebrities today—we feel a larger loss, that the collective world has lost something.

Suicide was once viewed as noble and heroic. In ancient Greece and Rome, forced suicide was a common form of execution, reserved mostly for aristocrats sentenced to death. Victims would ingest hemlock or fall on their swords. Offenders who knew they'd be punished harshly for breaking the law often took their own lives so their families could retain their property rights and belongings. Otherwise, the government could win ownership. As Christianity became the dominant religion in the Roman Empire, society's views on suicide gradually changed. By the sixth century, the once honorable act became a punishable sin. Christian burial laws soon stated that if you wanted a welcoming invitation to heaven, suicide was a big no-no. In AD 693, a mere attempt at suicide was seen as a crime, and one was harshly reprimanded for it, excommunicated from the church and subject to civil consequences.

The Romantics of the late eighteenth and early nineteenth centuries—think Byron and Keats—are to blame for glorifying and beautifying suicide. One of the first notables was seventeen-year-old British poet Thomas Chatterton, who, in 1770, tore his literary work into fragments before ingesting arsenic. A few years later he gained iconic notoriety as an unacknowledged genius for the Romantics. And during the French Revolution, suicide ceased to be a crime in many European countries.

As times changed, so did our opinions. Suicide has been looked at philosophically, ethically, religiously, and legally from antiquity into the nineteenth century. The 1800s brought with it a sociological/statistical inquiry paired with a psychological examination—posing the question "Why would someone kill himself?"—while the pill-popping Polly Prozacs of the 1970s gave us the birth of "antipsychosis" drugs such as Thorazine, and with it, the concept of a biochemical imbalance that has breathed new life into the analysis and prevention of suicidal tendencies and behaviors.

An estimated 75 percent of successful and would-be suicides give warning signs of their intensions. Many are driven to the act because their red flags or cries for help go ignored or unnoticed, which means at a certain point, someone will kill themselves in order to prove her seriousness.

"Very often, suicide is about control," says Dr. Edwin S. Shneidman, a leading suicidologist who has published close to two papers and twenty books on the topic. "It's the only time you can control death. You can control the exact hour, minute, day, and date. You call the shots. *Nature can't do this to me; I'll do it to me.*" Suicide also violates the basic rule in life: don't do something you can't come back from. "We want to comfort ourselves against the coldness of loneliness, and suicide is a surrender to that method of controlling one's own fate," Schneidman adds.

In 1999, concern for the number of suicides that were occurring was serious enough for the surgeon general to issue a "Call to Action to Prevent Suicide," defining it as a "public health hazard." And according to the National Center for Health Statistics, 90 percent of all people who successfully commit suicide have a diagnosable psychiatric disorder—such as depression—at the time of their death, nearly the same percentage as the people covered in this book. And yet depression alone will not trigger a suicidal act. Twice as many women as men experience depression, but men are four times more likely than women to commit suicide. It's important to consider drug addiction issues, too. Often, our icons suffered from both depression and addiction.

For as many methods and names as suicide has, just as many theories have surfaced as to why people kill themselves. In 1897, Emile Durkheim's classic *Suicide: A Study in Sociology* was published. This

French sociologist claimed that every suicide could be classified into four types: egoistic, altruistic, anomic, and fatalistic. He found that single people were more likely to kill themselves than married folks. Protestants were more likely to die from a self-killing than Catholics. And urbanites were at a higher risk than ruralists. Freudian analyst Karl Menninger said that there are three components to suicide: the wish to kill, the wish to be killed, and the wish to die.

Dr. Shneidman says that a suicide will happen when three factors collide. First, a person reaches his psychological threshold for pain—which results in his inability to envision any escape other than death. Next is easy access to life-ending tools—say, a gun, a knife, pills. Lastly, the person enters into "perturbation," an agitated state where he feels his discomfort and anxiety are intolerable. If these conditions coincide, a suicide attempt is inevitable. To reduce the threat, at least one of the three conditions needs to be reduced.

Thomas Joiner, a Florida State University Bright-Burton professor of psychology and the author of *Why People Die by Suicide* (Harvard University Press), adds another idea. He believes that those who kill themselves want to die, and have learned to overcome their instinct for self-preservation. Their desire for death is composed of two psychological states: a perception of being a burden to others and a feeling of not belonging. On their own, these states produce a desire for death, but together they create a need that can be life-threatening, especially when combined with an ability to enact self-injury.

But why a person commits or attempts suicide still remains an enigma under heated debate. The truth is, no one really knows. There isn't a single cause to which suicide can be directly attributed. There's no list of exact rules or criteria: Socrates, though forced, chose suicide on principle. Freud killed himself because of excessive pain, whereas mental anguish or intolerable circumstances might have driven Alan Turing to end his life. Hitler swallowed poison and then shot himself because of his grandiose, unrecoverable failure. Mental illness claimed the life of Virginia Woolf. And Kurt Cobain died when achievement brought him nothing but emptiness. Drug addiction, broken heart, loneliness, abandonment . . . and on and on.

What we do have are demographics and statistics. Age range and occupations. Methods and motivations. Facts and figures. While men are quick to reach for a gun, women gravitate toward overdosing or cutting themselves. Half of all people who commit suicide see a physician within a month of their fatal act. Parents, obese men, and pregnant women are less likely to commit suicide than anyone else. Divorced men commit suicide 400 percent more than women. And if you're a sixty-five-year-old male or older, in poor health, divorced, or have lost a loved one, and are living in a metropolitan area, you're part of the highest suicide group.

For some, there is planning—which brings momentary relief. It's the well-crafted, highly organized strategy that allows a suicide to write the notes and divide up her belongings, that gives her time to say her proper farewells—a temporary distraction from her inner agony as she maps out the best arrangement for how she will end it.

For others, self-killing is impulsive. Studies have shown that many attempters, 25 to 40 percent, are suicidal for only a short time. For some, it is as little as an hour. They awake one day without giving a thought to ending their life, and by evening they are ready to bid it adieu.

One person attempts suicide every thirty-four seconds. And one death by suicide occurs for every twenty-five attempted. Those seriously suicidal have been so for a long time, with the thought of how and when being a consuming rather than a momentary one. For the impulsive—who will often choose jumping or a gun as a means to their end—studies reveal that adding obstacles to their method slows down their irrational impulse, at which point the life-ending thoughts may pass. That's why many people lobby for barricades to be built on bridges, for guns to be locked up unloaded, and for medication normally stored in cabinets and dresser drawers to be removed.

Lastly, there's location and method, which are as specific and personal as the act itself. For the impulsive, this could mean a bridge. For the control freak, perhaps it's his home, surrounded by the comfort of prized possessions. Others select the quiet solitude of a hotel, where their act can take place uninterrupted and unnoticed, no mess for a family or friends to clean up. For some, such as actress Peg Entwistle, who jumped off the Hollywoodland sign, the method of suicide is symbolic, adding defiant significance to their act: the use of a father's gun, the swallowing of a mother's pills, the killing of a lover while on vacation and the subsequent shooting of one's self. Some leave notes; others surround themselves with press clippings, photos, and fan mail. And some just disappear, leaving their death more of a mystery while removing the negative stigma suicide can bring; without proof, anything is possible. Without hard evidence, nothing is written in stone.

Though suicide ends a life, it ironically keeps that person's life story alive. Regardless of method or location, planned or impulsive, the suicide's inner torment and utter anguish haunts our culture for decades.

DID YOU KNOW: It takes about ninety seconds to pass out and four minutes to die if you put a plastic bag over your head. Even more unsettling, the bag doesn't need to be tied at the bottom for the lungs to be deprived of oxygen. And forensics has no way of detecting trace evidence of the bag if you remove it from the victim.

THE TOP THREE SUICIDE METHODS		
Method	Males (%)	Females (%)
Gunshot	57	32
Suffocation	23	20
Poisoning	13	38

COPYCAT: Studies have found that when the suicide of a celebrity or political figure happens, a copycat effect is 14.3 times more likely to occur. Highly publicized stories increase the U.S. national suicide rate by 2.51 percent during that month of media coverage. Specialists attribute this growth to a greater degree of identification, and the following thinking: "If a celebrity, with all her fame, popularity, and fortune can't endure life, why should I?"

THE WERTHER EFFECT: In Goethe's 1774 novel, *Die Leiden des jungen Werther (The Sorrows of Young Werther)*, the novel's hero shoots himself after an ill-fated love affair. Many young men who read the book chose to end their lives by shooting themselves. The book was banned in some places, but the epidemic had already spread. This is one of the earliest documented cases of copycat suicide, and the reason why such suicides are often referred to as the "Werther effect," which was coined by David Phillips in 1974.

The first documented suicide note: "The Dispute with His Soul of One Who Is Tired of Life," translated in 1896

Two

The Art of the Suicide Note

The suicide note. A collection of words written impulsively in a crazed frenzy, or carefully, thoughtfully agonized over, so each word fits and flows seamlessly. Highly choreographed, overly manipulated, driven by madness, or calmly articulated—it doesn't matter. Each note is the same, each note is different—a last word leaving no room for rebuttal. Suicide notes are meant to explain, evoke sympathy, provide understanding, answer questions, or create new ones. They beg for forgiveness, confess deep, dark secrets, or attempt to hide things. Some point fingers, sharing the truth and thus setting off a spree of investigations.

Suicide letters have been addressed to siblings, husbands, wives, lovers, bosses, friends, fans, mentors, and enemies. Some suicides write a single document, others individualize each one, leaving a neatly piled stack of notes looking like paid bills or party invitations ready to be delivered. They are painstakingly written on parchment paper, hammered out on a typewriter, or scribbled onto a Post-it as in the case of singer-songwriter Elliott Smith's note, written in marker, pencil, crayon, and lipstick. Jules Pascin, a Bulgarian painter, wrote on a wall in his own blood, "Lucy, Pardonnez-moi," before hanging himself.

Some opt to use a photo, a visual substitution for the words they wish to communicate but can't seem to find. *New Yorker* cartoonist Ralph Barton shot himself in bed, his massive copy of *Gray's Anatomy* purposely opened to a picture of the human heart. Others add drama by videotaping their last wishes. Some send their final thoughts into the great unknown, relying on the Internet and e-mail to deliver their news. Whether carried by horse and buggy, hand-delivered, or sent by the postal service, these last letters find a way to their intended reader.

Suicide notes have even been mailed to loved ones ahead of time. Hunter S. Thompson sent one to his wife four days before shooting himself. They've also been purposely placed among one's belongings. Ian Curtis's note was found on his turntable, on top of his copy of Iggy Pop's album *The Idiot*. Or they've been left in someone's clothing—Sid Vicious's mother found her son's note shoved in the top pocket of his favorite leather jacket. They've been discovered on the suicide's body, as messages on an answering machine, or left in a handbag, as Peg Entwistle's was.

Though most suicide notes average 120 to 160 words, some souls have been able to muster only one word. Others, such as Cesare Pavese, an Italian poet who died from an overdose of barbiturates in 1950, wrote voluminous diaries. Some, such as German poet Paul Célan, simply underlined a sentence found in a book.

In many cases the letters are a puzzle, decipherable only to the people they're intended for. Hunter Thompson compared his life to sports, stating, "Football Season is Over." Others write in verse. The Russian poet Sergei Esenin hanged himself with his bathrobe belt on December 28, 1925, after writing a poem the day before in his own blood.

> Goodbye, my friend, goodbye
> My love, you are in my heart.
> It was preordained we should part
> And be reunited by and by.
> Goodbye: no handshake to endure.
> Let's have no sadness—furrowed brow.
> There's nothing new in dying now
> Though living is no newer.

"A suicide note is a mystery within a cryptic situation," famed suicidologist, Dr. Shneidman, explains. "Notes are an enigma of the suicide itself. And before one can attempt to understand the note, one must first understand the suicide." Notes are often written by those more prolific. There's a range of extremes—from the most simple, to the most archaic and complicated. And one out of every four people who kill themselves leaves one.

"The note essentially is an effort to make a statement, offer an explanation for the act, and/or to say goodbye and sometimes even an attempt by the suicider to consider the feelings of the soon-to-be survivors," remarks psychotherapist Jonathan Alpert. "Some are more autobiographical, where the person gives a synopsis of their life as they are reflecting on it. Others are more of a creative expression, and others are revenge, which is usually motivated by anger. Those who leave nothing may not have the clarity of mind, energy, or emotional stability to do so, as it does take some amount of focus."

Researchers say that about one third of suicide notes contains a request, ranging from how their body should be cared for, to what the inscription on their tombstone should read, and even to how their belongings and financial assets should be divided.

"We don't know much about suicide notes because there are no studies that show a demographically different group," adds Dr. Shneidman. "The purpose is to rationalize, excuse, and explain the act itself. It's the *Please try to understand me*. It's also the empathy some are to convey, an arrow of affection of emotion and a way to underscore the ambivalence of the act itself." Reasons include guilt, anger, explanation, redemption, pity, loathing, sadness, loss, illness, sanity, and insanity.

Some write simple instructions. Sylvia Plath had one request: "Please call Dr. Horder," her psychiatrist. Dorothy Dandridge gave a note to her manager that read, "In case of my death, to whomever discovers it, don't remove anything I have on—scarf, gown or underwear. Cremate me right away. If I have anything, money, furniture, give to my mother Ruby Dandridge. She will know what to do."

Some write about their fear of the future. Virginia Woolf knew her inner darkness had returned. "I feel certain that I'm going mad again. I feel we can't go thru another of those terrible times. And I shan't recover this time. I begin to hear voices, and I can't concentrate. So I am doing what seems the best thing to do." In a letter to his imaginary childhood friend, Boddah, Kurt Cobain lamented his loss of passion and not being able to fake it anymore.

Some suicide notes are meant for a single pair of eyes; others reach out to the masses. Ralph Barton mailed his suicide note to the *New York Times*, hoping the newspaper would print his last words.

Suicide notes have been found folded into perfect squares, left in pristine condition, or shredded into bits. Former deputy White House counsel Vince Foster tore his suicide note into twenty-seven pieces before dropping them into the bottom of his briefcase.

Some suicide notes are found in date books—literally, a last scheduled appointment. Diane Arbus wrote "The last supper" as her final to-do for July 26. Spalding Gray's last journal entry, dated December 2003, was written to his wife, Kathie: "It's an old story you've heard over and over. My life is coming to an end. Everything is in my head now. My timing is off. In the last two years I've had at least ten therapists and all those shock treatments. Suicide is a viable alternative for me instead of going to an institution. I don't want an audience. I don't want anyone to see me slip into the water."

Suicide notes are written sober or intoxicated. Some writers of suicide notes are coming off a bender, while others are slowly slipping into one. Sometimes the notes are as baffling as the acts of self-destruction that follow. One thing is for sure, each is created when the writer was at his psychological worst, his logic distorted.

Prior to the eighteenth century, very few suicide notes were written. Some specialists attribute the reason for this to lack of literacy. In the 1700s, newspapers came to the forefront, supplying eager readers with rich information. Often suicide notes were read aloud on a stage in front of an audience. The intimacy removed, many felt these epistles were written to be published, read to their peers in an attempt for last moments of glory.

In early times, these documents were inscribed to absolve the suicide's family of punishment. Lewis Kennedy, the Duke of Bedford's gardener, who slit his throat in 1743, addressed his suicide note to "any random Christian and Courteous Reader." Specialists say Kennedy's note was specifically crafted so his family would not be blamed financially or incriminated for his wrongful act.

According to psychiatrist Chris Thomas, the first suicide note dates back almost four thousand years, to between 2000 and 1740 BC, authored by an ancient Egyptian known as "the Eloquent Peasant" and commissioned by King Meri-ka-re. The document was titled "The Dispute with His Soul of One Who Is Tired of Life." Egyptologist Adolf Erman translated the papyrus in 1896. The poem was intended to discourage the king's subjects from taking their own lives. He had written several poems in hieroglyphs on papyrus. Today the scripture resides in Berlin's Egyptian Museum and Papyrus Collection. Having analyzed it, specialists believe that it reveals the mind of a deeply depressed man: "To whom do I speak today? I am laden with misery, And lack a trusty friend." Later, he continues with "Death is before me today, As a well-trodden path, As when a man returneth from the war unto his house."

Often suicide notes intended to remain private are made public—Hunter Thompson's was published in *Rolling Stone*. Others find their way into obituaries, glossy gossip rags, and nightly news shows.

Some suicides, such as actor David Strickland or artist Mark Rothko, leave nothing at all. Just a trail of unanswered questions leading to an unsolved puzzle.

In the past fifteen years, suicide notes have been examined for their syntactical and graphological characteristics, their semantic similarities, and their organization of thought. Handwriting specialists and graphologists have become popular over the years as suicides have become more complicated. Four or five different handwriting experts were hired to deconstruct and scrutinize the authenticity of Kurt Cobain's suicide note.

What's most bizarre is the similarity among suicide notes. Whether written in 1800 or 2000, their authors women or men, foreign or American, adolescent or elderly, these notes deviate little. Each is a last, desperate attempt to share and explain, communicate, and of course say goodbye. To leave something tangible behind after they've taken their lives. There's also a bit of immortality—the words, the intent, even the person, has forever been encapsulated in a note that, regardless of the means by which it was written or delivered, creates for him purpose and permanency.

UNEARTHED: Monday is the most common day to kill yourself, whereas Saturday is the least popular. The idea that most suicides take place over Thanksgiving, Christmas, and New Year's Eve is a myth. Most suicides happen during the springtime. The thinking is this: people expect the winter months to be more depressing; so when cheerful summer appears, and things have not gotten better, they feel a greater sense of despair, and thus take their own lives.

Three
Authors

Seneca once wrote that the road to freedom could be found in every vein of one's body. Seneca's adage could allude to the basic rule of "write what you know." In AD 65, the emperor Nero forced Seneca to kill himself by severing the arteries in his arms, knees, and legs. Other authors have written what they know: in their scriptures, poems, novels, and memoirs, putting blood, heart, and soul onto the page.

Nothing could be truer for Virginia Woolf, Ernest Hemingway, Hunter S. Thompson, Sylvia Plath, and Anne Sexton. Their work was confessional, descriptive, and edgy. They challenged and confronted, often pulling the reader into their inner world of darkness, despair, and depression. And between 1941 and 2005, suicide claimed the lives of these five, among the twentieth century's most influential and recognized writers.

Like musicians who put pen to paper to create their lyrics, writers are under similar scrutiny. And when they commit suicide, their words are easily taken out of context by the overly analytical or the eager critic playing detective and positioning their works under a magnifying glass to understand their deaths. Whether scribbled quickly on a bar napkin, written down on a scrap of paper, or listed thoughtfully in a diary or journal, the feelings and thoughts of a writer during his lifetime are not always fair game for citing as evidence of suicidal tendencies: "See, it's right here. He used the word *suicide* nine times in his last poem."

In 2001, James Pennebaker, a University of Texas psychology professor, and Shannon Wiltsey Stirman, a graduate student at the University of Pennsylvania, conducted a study of poets who'd killed themselves, and whether their writing foreshadowed their demise. The study looked at the work of John Berryman, Hart Crane, Sergei Esenin, Adam L. Gordon, Randall Jarrell, Vladimir Mayakovsky, Sylvia Plath, Sarah Teasdale, and Anne Sexton. The researchers found that poets who committed suicide used many more first-person singular references (such as *I, me,* and *my*) and fewer first-person plural words (*we, us,* and *our*) than did non-suicidal poets.

"Issues of identity, isolation, and connection to others is revealed in pronoun usage," Pennebaker said in an interview published in *Wired* magazine. "One of the most telling words of all is the word 'I.' People who are suicidal or depressed use 'I' at much, much higher rates, and there's also a corresponding drop in references to other people." These writers embraced their sadness and misery while analyzing themselves from a more distant perspective.

Pennebaker's previous research found that suicide rates were far greater among poets than among other literary figures and the general public. Not surprising, since poets seem more prone to depression, bipolar disorder/manic depression, and other related illnesses.

Considered the founding father of confessional poetry, John Berryman had a Jekyll-and-Hyde personality when it came to drinking. An alcoholic by his twenties, the *Dream Songs* author ended his life in 1972 when he drunkenly headed for the Washington Avenue Bridge in Minneapolis, Minnesota. "I am a nuisance," he wrote on January 5, in a suicide note he left on his kitchen table. Then he walked out the door, only to return moments later to scribble a long poem about his inability to follow through with his death wish and being fired from his job. Frustrated, he crossed out the poem and tossed it in the trash. Two days later, he appeared again at the same bridge and jumped.

In the mid- to late nineteenth century, three significant suicides occurred: Gérard de Nerval, a French translator and favorite among German Romantics, hanged himself with apron string in Paris in 1855; Constance F. Woolson, an American novelist living in Venice, jumped from her window in 1894; and in 1896, José Asunción Silva, a thirty-one-year-old Colombian poet, retired to his room and fired a revolver into his heart.

A slew of others followed, most notably Pulitzer Prize–winning poet Sara Teasdale, who spent most of her twenties fixated on death, and in 1933 finally killed herself. Suffering from financial woes thanks to the Great Depression, a divorce from her husband, and the death of a close friend, Teasdale decided to join him by crawling into a warm bubble bath in her Fifth Avenue Manhattan apartment and overdosing on sleeping pills.

Virginia and Leonard Woolf's home, Monk's House in the village of Rodmell

Virginia Woolf

BORN: January 25, 1882, London

DIED: March 28, 1941, Lewes, East Sussex

AGE: 59

METHOD: Drowning

DISCOVERED BY: Children playing by the water

FUNERAL: Unknown

FINAL RESTING PLACE: She was cremated at Brighton on April 21. Her ashes were scattered under one of the elms at her home, Monk's House, in the village of Rodmell. A quote picked out by her husband, Leonard Woolf, from her novel *The Waves* resides under her bust.

I meant to write about death, only life came breaking in as usual.

FROM HER DIARY, FEBRUARY 17, 1922

At 11:30 AM, the weather in Sussex, England, is brisk, the sun shining. The large stones are smooth in her hands. Solid and heavy in her pockets. They bulge from her coat. Though she found herself in this exact position days ago, standing by the river, ready to end her life, she failed. She returned home drenched, shivering from the cold. But today she knows what to do. Today she has the rocks.

Leaving her walking stick by the bank, she proceeds slowly, purposefully into the River Ouse. The water is cold and numbing. She moves farther and farther into the high tide until she is part of the river. Until the voices in her head are silenced. She no longer hears a chorus of birds singing in Greek or King Edward VII spewing obscenities. There will be no more migraines, insomnia, shaking of hands, loss of words, or feelings of inadequacy. The water enters her mouth and lungs and engulfs her airways. Until she is unable to breathe.

Virginia Woolf spent most of her life in one of two states: writing or fighting a bipolar disorder/manic depression that went undiagnosed until after she'd killed herself. Regarded as a feminist and modernist, she penned many novels and essays that closely mirrored her life, her madness, and her frequent contemplation of suicide.

She was often a dichotomy of characteristics: laughing and talking frantically as witticisms spewed from her mouth, or shy, withdrawn, awkward, and reserved. One moment she could be kind and sweet; the next, malicious and snotty. The warning signs were many. It would start with a growing nervousness and fear of people. Her heart would beat quickly, her mind race with thoughts and voices. At her worst, she'd stop eating and working. She became violent and moody, uttering gibberish. For a woman who was clever, who depended on her love of words and the exactness of her observations, her illness was like a death sentence.

From reading her countless journals, it is easy to see that she was aware when it was happening. After therapy, shock treatments, stays at convalescent hospitals, and medication failed her, she would depend on four round-the-clock nurses.

During her lifetime she had numerous breakdowns. The first came shortly after her mother died, when Virginia was thirteen, and lasted for more than six months. The second occurred after the death of her father, nine years later, which was followed by a suicide attempt in the summer of 1904. She was staying at a friend's home and, though she had constant care, was able to crawl out a window and jump. The house was only one story high, and her attempt failed. More madness came when she was twenty-eight, and then again at age thirty. She would hallucinate, hear voices, and speak incoherently. Mental illness was rampant in her family. Though she grew up in an intellectual home, it was filled with tension. Her father and three siblings each dealt with emotional instability throughout their lives. Her sister Vanessa fell into a two-year depression in her thirties after a miscarriage, closely reflecting Virginia's symptoms. Her first cousin on her father's side developed severe mania in his twenties, and died within a few years in an asylum. And though she went to the best therapists, it was still the early 1900s. Psychiatry hadn't caught up to the mental disorders that later emerged in Virginia, and so she was often misdiagnosed.

She met her husband, Leonard Woolf, a writer, through the Bloomsbury Group, which comprised important twentieth-century authors who lived in or near London. Instantly smitten, Leonard proposed to her three different times. They married in 1912. Two years later, she wrote her first novel, *The Voyage Out*. She was originally pleased with the project, but it took only a few weeks for her opinion to change and for her to view the work as unworthy. A second suicide attempt happened when Leonard went to visit Virginia's doctor. While he was gone, she swallowed one hundred grains of Veronal, which is part of the barbiturate family and acts as a sleep aid. By chance, a surgeon from St. Bartholomew's Hospital was lodging at the same hotel as the Woolfs, in Brunswick Square. He ran to the hospital, grabbed a stomach pump, and saved her life.

While her family grew unsympathetic of Virginia's mental state, often referring to her as "the goat," it was Leonard who took care of her for the twenty years of their relationship. Though he didn't know she had a mental disorder when he married her, he not only accepted the role of caretaker, but excelled at it. Married to Virginia only a year when her third breakdown occurred, he became her memory, recording in journals her every emotion, feeling, and action. He kept a pocket diary to record her moods, creating code names made up of Tamil and Sinhalese characters.

In February of 1915, she relapsed, becoming completely incoherent. "One morning she was having breakfast in bed and I was talking to her when without warning she became violently excited and distressed," Leonard recalled in his autobiography *Beginning Again: An Autobiography of the Years 1911 to 1918*. "She thought mother was in the room, and began to talk to her. It was the beginning of the terrifying second stage of her mental breakdown."

This was followed by two or three days of unrelenting gibberish.

Virginia's relationship with her husband was more of a deep, intense friendship than the traditional sexual relationship sought by most couples. Sexually abused by a half-brother when she was five or six, and again by another brother as a teenager, she found solace in women and had several lesbian affairs, most notably with poet-novelist Vita Sackville-West.

Virginia smoked cheroot after dinner, and loved jokes and gossip. She was ultra-smart and humorous, with a fondness for children, though she and Leonard never had any because of her illness. She was considered unattractive, with sharp features and a thin, lanky frame, and she always appeared frail.

From mid-1940 through the New Year, the ramifications from the Second World War started to affect southern England. Air raids and the mounting threat of Nazi invasion worsened Virginia's unstable state. She was concerned that Leonard, a Jew, would be taken away, that they would be separated. They decided that should a German attack happen, they would shut their garage door and commit suicide together with a lethal dose of morphine.

During calm periods, when her head was quiet, she wrote, somehow managing to juggle three writing projects simultaneously. The last work she completed was *Between the Acts*, which she sent to publisher John Lehmann, who quickly noticed a higher-than-normal number of typing errors and smudges—Virginia's attempts to correct her mistakes.

"The spellings were more eccentric, more irregular than in any typescript of hers I had seen before," he once mentioned. "Each page was splashed with corrections, in a way that suggested that the hand that had made them had been governed by a high voltage electric current."

As her depression worsened, the excitement for this final novel drained from her. She hated it, thought it pitiful and unpublishable. A few days before her death, she sent Lehmann a letter insisting that the work was trivial and not ready for publication. Leonard sent an accompanying note explaining that his wife was on the verge of a breakdown.

A day before her suicide, Virginia saw Dr. Octavia Wilberforce, a therapist she'd met with before. Dr. Wilberforce was a newcomer to the field, and easily seduced by Virginia's brilliance and artistic personality. Rather than recognize the clear symptoms of depression, she informed Leonard that a nurse would not be necessary, that Virginia seemed positive.

Twenty-four hours later she was dead.

The site of Virginia's suicide, the River Ouse

The day of her suicide she rushed out of the house, hoping not to be seen, but was spotted by their housekeeper, Louise. At 1:00 PM, Louise rang the bell indicating that lunch was ready. Leonard left his workshop and headed upstairs to the sitting room to listen to the radio, where he found Virginia's handwritten notes on the mantel. It's unclear if he opened them, but he instinctively knew what they were and rushed downstairs, yelling for Louise. As he ran to the river, she went for help, and the gardener rushed to fetch the police. A mile from the bridge, Leonard found Virginia's walking stick. A team of people searched the river with rope and tackle until it grew dark. Everyone came up empty-handed.

It took three weeks for her bloated body to be spotted. A group of children stumbled upon it accidentally. An inquest was held the following day at Newhaven, where the verdict, in the standard phrase of the time, was "suicide while the balance of her mind was disturbed." After taking care of Virginia for two decades, Leonard's last responsibilities were to identify her body and have her cremated.

Virginia's note to her sister stated how much Vanessa and her children had meant to her, how much she loved them, and for them to please help Leonard through this loss: "Leonard has been so astonishingly good, every day, always; I can't imagine that anyone could have done more than he has . . . will you assure him of this."

Two letters were addressed to Leonard, one written the day of her suicide; the other had been penned a few days earlier, when he thought perhaps she tried to drown herself for the first time. (She'd come home dripping wet, shivering from more than just the rain. He asked what had happened and she told him she'd slipped into the river.) The earlier note asked him to destroy all her papers. It also acknowledged the burden that suicide inflicts on those left behind.

The second note, below, was written that morning.

Dearest, I feel certain I am going mad again. I feel we can't go through another of those terrible times. And I shan't recover this time. I begin to hear voices, and I can't concentrate. So I am doing what seems the best thing to do. You have given me the greatest possible happiness. You have been in every way all that anyone could be. I don't think two people could have been happier till this terrible disease came. I can't fight any longer. I know that I am spoiling your life, that without me you could work. And you will I know. You see I can't even write this properly. I can't read. What I want to say is I owe all the happiness of my life to you. You have been entirely patient with me and incredibly good. I want to say that—everybody knows it. If anybody could have saved me it would have been you. Everything has gone from me but the certainty of your goodness. I can't go on spoiling your life any longer. I don't think two people could have been happier than we have been.

In a strange twist, Leonard received an anonymous letter informing him that the *Sunday Times* had misquoted her suicide note. The misquoted note read, "I feel I can not go on any longer in these terrible times," which the anonymous writer said made it sound as if the war had seduced her into killing herself. Leonard wrote to the newspaper asking for a correction, which they obliged.

UNEARTHED: With all the aliments Virginia had—and there were many—she may also have suffered from catabythismomania, a morbid impulse to commit suicide by drowning.

Virginia, 1928

DROWNING: The least used method—accounting for only 1.1 percent, or approximately 375 deaths, per year—drowning is also the most mentally demanding and painfully agonizing one. The suicide needs to focus on being still and not breathing while suffocation occurs underwater. Weights are useful to combat the body's instinct to convulse and try to rise to the surface. The average person loses consciousness within three minutes of being underwater. Brain damage usually happens after one is deprived of oxygen for five or six minutes. Women prefer to do it in the ocean—favoring large bodies of water—and bathtubs, while men choose rivers, ditches, and lakes. Swimming pools are rarely used.

WHAT HAPPENS: When water enters the throat, it triggers a spasm that blocks the windpipe and prevents water from entering the lungs. With no air coming in, and the lungs being depleted of air, the person dies. At a depth of one hundred feet, the lungs compress to a quarter of their original volume.

CAREER HIGHLIGHTS: Each of Virginia's homes and their locations play a prominent part in her work. *Mrs. Dalloway* was written in London. The title character in *Orlando* was born in Knole, where Virginia's lover Vita Sackville-West resided. Cambridge gave us *A Room of One's Own*; and Monk's House is where Woolf wrote her final tome, *Between the Acts*. Aside from her many published works, she was also a book reviewer and essayist, contributing regularly to *The Times Literary Supplement* and *Cornhill*. She and Leonard founded the Hogarth Press, which introduced important new works of literature and published the twenty-four-volume edition of Freud's writings in English translation. Woolf is also known for writing *Mr. Bennett and Mrs. Brown*, *To the Lighthouse*, *The Waves*, and *Between the Acts*, among others.

Ernest Hemingway

BORN: July 21, 1899, Oak Park, Illinois

DIED: July 2, 1961, Ketchum, Idaho

AGE: 61

METHOD: Gunshot

DISCOVERED BY: His fourth wife, Mary

FUNERAL: His three sons, who had not been in the same room in fifteen years, stood next to Mary on one side of their father's grave, while three of his sisters stood on the other. Throngs of fans stood outside the graveyard perimeter.

FINAL RESTING PLACE: Buried between two towering pine trees in Ketchum Cemetery, in Ketchum, Idaho. A bust of Hemingway was placed on a stone pedestal at Trail Creek, outside of town, five years later.

> *Happiness in intelligent people is the rarest thing I know.*
> ERNEST HEMINGWAY

"Look, this is how I'm going to do it," he said, joking to the friends he was hosting in his Cuba home known as Lookout Farm. He placed the double barrels of the twelve-gauge English shotgun in his mouth. The metal, cold and hard; his mouth warm and soft. He looked at his shaken guests and smiled cunningly, getting off on how uneasy he was making them, and removed the weapon from his mouth. "The palate is the softest part of the head," he added.

In a reenactment years later, Ernest Hemingway, dressed in a robe and pajamas, padded down the steps of his two-story house in Ketchum, Idaho, and, imitating a thief, stole a set of keys from the

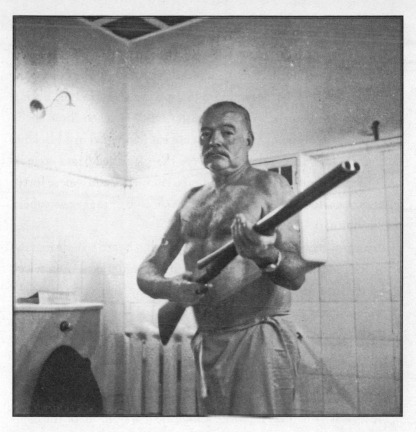

Ernest Hemingway with one of his beloved shotguns at the Finca Vigia, Cuba, 1954

kitchen windowsill, went down to the basement, and unlocked the gun rack to select his favorite weapon.

Considered one of the greatest, most important writers of his time, Hemingway was a six-foot-tall, burly bear of a man with a barrel chest and massive muscled arms. With a passion for danger and adventure, he was the epitome of a man's man. He was a boxer, a hunter, an expert in firearms—he was given his first shotgun at age ten. He fished off the coast of Cuba, journeyed to Paris and Spain, and followed the bullfighting circuit, running with the bulls in Pamplona. He somehow survived many near-death situations: blood poisoning on an African safari, a wounding by mortar shells in Italy during World War I, an escape in the Spanish Civil War when three shells entered his hotel room, a taxi accident that happened in a blackout during World War II. Then there were the plane accidents—one in 1954, while he and his wife were on safari in Africa, and a mere forty-eight hours later, a second one, which was so bad that newspapers ran his obituary.

He was rarely without a pipe in his mouth, a beard to hide his jaw, a bottle of whiskey by his side, or a gun in his hand. He had four wives and three sons. He lived hard, drank excessively, and wrote often. At twenty-seven he insisted on being called Papa.

A true bully, he was boastful, boorish, and belligerent. He picked fights in bars. He was severely competitive, which made him a sore winner and loser. A masterful storyteller, he often exaggerated his tales for shock value or to help prove how incredibly important and brave he was. Whatever he wrote about, be it war, food, or the complexities of man versus the world, he excelled.

A stubborn alcoholic during the last twenty years of his life, he consumed a quart of whiskey a day. He had many hospital stays, several at the Mayo Clinic for high blood pressure and a persistent case of hepatitis. He also received electroshock therapy, which prevented him from writing, a residual effect that added greatly to his mounting depression. Upon returning home from the hospital in 1961, he became paranoid, insisting that the government was following him. Feeling washed up and depleted, he would spend hours staring out the window at the cemetery across the river. For a voracious writer, at the end of his life he suffered from writer's block on top of his depression, and five hundred words a day was considered a monumental accomplishment for him.

In April of that year he became highly suicidal, and on the twenty-first, his fourth wife, Mary, found him standing in the corner of the living room clad in his red Italian bathrobe, which they had fondly dubbed "the Emperor's robe," a shotgun in his hand. The gun's breech was open, and on the windowsill nearby lay two shells. He'd written a note, which he refused to give to his wife, but he conceded to reading part of it. His doctor was coming at noon to take his blood pressure, so Mary soothed the beast for the next fifty minutes, assuring him that his writing ability would return, reminding him that he had children who would miss him, that she needed him desperately.

When his doctor arrived, he persuaded Hemingway to hand over the gun, then checked him into Sun Valley Hospital, where he was put under heavy sedation.

Another trip to the Mayo Clinic was scheduled. Upon returning home to pack for the longer hospital stay, and though accompanied by two friends, Hemingway was still able to race downstairs to his gun collection and reach one before anyone could catch him. As he placed the barrel in his mouth, it was wrested away by his friend.

Another hospital visit followed.

A second attempt to get him to the Mayo Clinic was punctuated by another escape when, at the airport, he broke free from his minders and walked purposely close to a plane's propellers. When this didn't bring him the result he intended, once inside his plane, shortly after takeoff, he tried to open the door and jump out.

Throughout his life Hemingway had odd habits and rituals. He'd lug around samples of his own urine. He never wore underwear and disliked bathing in water. Instead, he took sponge baths with rubbing alcohol. He would call death "Nada." One June, when a friend came to visit, he found Hemingway edgy and antsy, suffering from delusions and high blood pressure. He insisted he was going blind and becoming impotent.

It all came to an end in July 1961. Hemingway and his wife had been sleeping in separate rooms in their two-story bunker-block chalet. Hemingway unlocked the gun rack, selected his baby—a double-barreled Boss shotgun with a tight choke, from Abercrombie and Fitch—and climbed the stairs to the main floor. In his oak-paneled foyer he slipped two shells into the barrel, lowered the gun butt to the floor, bent down on his knees, pressed his mouth to the barrel, and pulled the trigger. Mary later reported that the sound was that of dresser drawers banging shut. Jarred awake, she ran downstairs to find her husband crumpled on the floor, surrounded by bits of brain, bone, and blood. The gun lay beside him with one chamber discharged.

The funeral was delayed a few days in order for his three sons to return home. Hemingway had been excommunicated by the Catholic Church after his divorce from his second wife, Pauline, and thus was denied a complete funeral mass. His family had hoped for a Catholic burial to show that his death wasn't a suicide. This created a huge uproar among the family members, the Church, and one priest, who said the Church accepted the ruling of the authorities, which stated that his wound had been self-inflicted. A compromise was reached: his funeral would be at the grave, not the church, but would be a Catholic one, albeit not High Mass.

During the reading of the will, his sons found out that they had

BEST OF ALL HE LOVED THE FALL
THE LEAVES YELLOW ON THE COTTONWOODS
LEAVES FLOATING ON THE TROUT STREAMS
AND ABOVE THE HILLS
THE HIGH BLUE WINDLESS SKIES
...NOW HE WILL BE A PART OF THEM FOREVER
ERNEST HEMINGWAY · IDAHO · 1939

Hemingway's grave at the Ketchum Cemetery, Ketchum, Idaho

been disinherited. On September 17, 1955, their father had written, "I have intentionally omitted to provide for my children now living or for any that may be born after this will has been executed, as I repose complete confidence in my beloved wife Mary to provide for them according to written instructions I have given her." Though he didn't have life insurance, Mary inherited his estate—worth $1.4 million—and was named his literary executor.

UNEARTHED: The leading method of killing yourself, a gunshot is easy, fast, painless, efficient, and accounts for approximately 52 percent of all suicides. "Pulling a Hemingway" and the "Hemingway solution" refer to killing yourself by placing a shotgun to the head. People often put shotguns or rifles in their mouths, since it's a stabilized position and provides a direct route to the brainstem. Since most people are right-handed, guns are also usually aimed at the right temple.

FAMILY AND MADNESS: If suicide is genetic and runs in the family, the Hemingway family tree is heavy with mental illness. At fifty-seven, his father, a physician who conditioned his son to "be afraid of nothing," shot himself with his father's Civil War pistol. Two of Hemingway's five siblings, Ursula and Leicester, also took their lives, as did his forty-one-year-old granddaughter Margaux, who overdosed on sedatives in 1996. His son Gregory, who changed his name to Gloria after a sex-change operation, died in police custody after being picked up in a drunken stupor.

CAREER HIGHLIGHTS: Hemingway was a lean, forceful, and crisp writer. Always searching for "one true sentence," he would write one, then another, and another, filling blue notebooks with pencil scrawling. He thought himself a spokesperson for the Lost Generation, the disillusioned post–World War I generation, and many would agree. He won the Pulitzer Prize in 1953, followed by the Nobel Prize in 1954. He also graced the cover of *Time* twice. *Islands in the Stream* and *The Garden of Eden*, though unfinished manuscripts, found their way to being published, along with many other of his works, after his death. The last book to be published from the grave was *True at First Light*. His last living published work, in 1960, was an article on bullfighting for *Life* magazine. Commissioned to write 10,000 words, but unable to edit or write cohesively, he had turned in a manuscript of over 120,000. Though all of his fifteen works were masterful, most notable among them were *The Sun Also Rises, A Farewell to Arms, For Whom the Bell Tolls*, and *The Old Man and the Sea*.

PUBLISHED POSTHUMOUSLY: *The Wild Years* (compilation, 1962); *A Moveable Feast* (memoir, 1964); *By-Line Ernest Hemingway: Selected Articles and Dispatches of Four Decades* (journalism, 1967); *Islands in the Stream* (novel, 1970); *The Nick Adams Stories* (1972); *88 Poems* (1979); *Ernest Hemingway Selected Letters: 1917–1961* (1981).

Hunter S. Thompson at his home in Wood Creek in September 1990

Hunter S. Thompson

BORN: July 18, 1937, Huntingville, Kentucky
DIED: February 20, 2005, Woody Creek, Colorado
AGE: 67
METHOD: Gunshot to the head
DISCOVERED BY: His son
FUNERAL: Among the celebrity guests at the memorial bash were

Jann S. Wenner, founder and editor of *Rolling Stone*; Senator John Kerry; former U.S. senator and Democratic presidential candidate George McGovern; *60 Minutes* correspondent Ed Bradley; TV interviewer Charlie Rose; country crooner Lyle Lovett; and actors Bill Murray, Jack Nicholson, Sean Penn, Josh Hartnett, and Harry Dean Stanton.

FINAL RESTING PLACE: Ashes scattered into the sky

> *Buy the ticket, take the ride.*
> FROM *FEAR AND LOATHING IN LAS VEGAS*

The $2 million mammoth cannon shaped like the Gonzo fist—Hunter S. Thompson's trademark—had been built. The fifteen-story tower, which peaked two feet taller than Miss Liberty, with its green peyote light in the middle, was a spectacle unto itself. It was packed with ten shells made up of gunpowder and Hunter's ashes. His wife, Anita, had inscribed "I love you" on each shell before they were driven away by an armored car and packed into the gun. It was only part of the attraction that loomed over Owl Farm, Hunter's home in Woody Creek, Colorado. A menagerie of odd objects—stuffed peacocks and Chinese gongs and other assorted "Hunter" artifacts—were scattered about. His apple-red convertible, containing blow-up dolls, was parked on a nearby knoll, and photos of his favorite authors—Hemingway, Faulkner, Conrad, Twain, and Fitzgerald—displayed alongside pictures of the excessive writer himself. Construction for the memorial had begun two months earlier, in June, and the end result promised to be an elaborate and bizarre funeral.

Invitations to the affair, which featured an emblem of a silver foil dagger topped by a double-thumbed fist, with a street value of a Willy Wonka golden ticket, had been sent to three hundred plus handpicked people. At 7:00 PM, shuttle buses pulled out of the Hotel Jerome—a favorite haunt of Hunter's since the mid-1960s; he'd used the hotel's J-Bar as a veritable office. The shuttle bus passengers waved and smiled to the wall of cheering fans lining the route, and held up bottles of wine and liquor in a toast, paying homage to their hard-core journalistic friend. Among the mass of admirers were those dressed as their hero: Dunhill cigarettes dangling from their lips, Tilley hats hiding their eyes, white Converse sneakers on their feet. Only Hunter could have planned such an ornate show, and in essence, he had.

Originally conceived in 1977, the funeral celebration was the creation of Ralph Steadman, Hunter's illustrator sidekick, whom he'd asked to create a blueprint for a Gonzo Fist Ceremony. The idea for his memorial had been brewing in his warped mind for years, to the point where it had become an obsession. It was his "one true wish," he often called it.

In his vision, he saw a massive cannon that would shoot his ashes, spreading them over his friends and family while Bob Dylan's "Mr. Tambourine Man" boomed from above. An eight-piece Japanese drum band and a Buddhist reading would be part of the ceremony. On Saturday, August 20, 2005, it all became a reality—thanks to longtime friend Johnny Depp, who paid for the entire event.

Hunter's son, Juan Thompson, requested short testimonials from family and friends, and Anita, Hunter's second wife, sobbed her way through Coleridge's epic poem *Kubla Khan*. Lyle Lovett sang, as did Depp and Hunter's brother. "Spirit in the Sky" blasted from loudspeakers. The Heart Sutra was read in Tibetan, followed by a performance by a troupe of Japanese drummers. Mint juleps were replaced with champagne flutes and raw oysters, and Gonzo-emblazoned chocolates were part of the fare. At 8:46 PM, fireworks illuminated the sky, followed by the firing of the cannon. Hunter's ashes were spewed into the air, and then fell like specks of gray snow. As requested, Dylan's "Mr. Tambourine Man" played to the silent crowd. For a goodbye lullaby, "My Old Kentucky Home" was one of the last songs of the evening.

The event was a near-perfect replica of the big picture. And though Hunter was there in spirit, he'd missed the biggest party of his life.

Six months earlier, at 5:42 PM, he was sitting on a stool at his desk, located in the kitchen of his home, dressed in a red-and-black-striped shirt, worn inside out, sweat pants, and slippers. Hunter placed his glasses on his face and phoned his wife, who was at the gym, and with his son, daughter-in-law, and grandson in the adjacent room, he raised the forty-five-caliber gun to his mouth and pulled the trigger, leaving Anita on the other end of the line to be part of the experience. A piece of stationery from the Fourth Amendment Foundation (established to defend victims of unwarranted search and seizure) was inserted in his typewriter. He had typed only one word on the blank page: *counselor*.

Hunter's son thought the sound from the gun was from a fallen book, but checked on his father anyway and found him sitting at his typewriter, lifeless and pale, his head forward, chin resting on his chest, his hands in his lap. By his right foot was the bloody gun and its case. Another gun was on a shelf above his chair. Blood pooled on the floor. It dripped from his mouth. It was splattered on the kitchen counter and stove.

By the time the police arrived, Hunter's son was outside, a gun in his hands. He had fired three shots, symbolizing the passing of a great man. Afterward he went into the kitchen and draped a golden orange scarf over his father's shoulders, which his wife had given to Hunter the night before. The police felt for a pulse—there was none. They then searched for the slug and found it in the stove's hood.

Four days before his death, Hunter had sent a suicide note scrawled in black marker to his wife titled "Football Season Is Over." She received it the day he died.

No More Games. No More Bombs. No More Walking. No
More Fun. No More Swimming. 67. That is 17 years past 50. 17
more than I needed or wanted. Boring. I am always bitchy. No
Fun—for anybody. 67. You are getting Greedy. Act your old age.
Relax—This won't hurt.

At the bottom of the page he drew a heart.

Hunter was notorious for many things: his bad-boy persona, his passion for guns, his drug addiction. His outspoken antics and Gonzo writing style earned him fame and a cult following and millions of fans. It also earned him trouble with the law, trouble with editors, and a place in journalistic history. His husky, throaty, froglike voice was as unique as the outfits he sported. He was rarely seen without a Dunhill in a cigarette holder hanging from his lips, his aviator sunglasses, or his trademark hat, shorts, and sneakers. He was hard on his system and on his interviewees. His life was one long acid trip—filled with political activism, controversy, and excessiveness. The ultimate outlaw and icon of counterculture, he'd drive 110 miles an hour with the headlights off, drunk and high at 3:00 AM, a bottle of liquor in one hand, a cigarette in the other, a handgun between his knees. He never apologized for his behavior, and few people asked him to.

Hunter seemed to have been born this way. He missed his high school graduation because a soda machine had eaten his quarter and he'd used a shotgun to get even with it. He was arrested and thrown in jail for thirty days.

After a stint in the air force, he switched career paths in the late 1950s and ended up working as a sports editor for a local newspaper in Pennsylvania before moving to New York City. There he briefly worked at *Time* magazine, earning fifty-one dollars a week as a copy boy before being fired for insubordination. This became a running theme for Hunter, and he would lose job after job for his odd and unprofessional behavior. In 1960 he and his first wife moved to San Juan, Puerto Rico, so he could take a job with the magazine *El Sportivo*. There he started writing a novel while working as a security guard. He wrote a story about this experience, which got him fired, but got his writing noticed. Other work came from the *National Observer, The New York Times Magazine, Esquire*, and *The Nation*— which brought him his first major article, on his experience with the California-based Hells Angels motorcycle gang. In 1966 Random House published a book about his experiences following these hard-core anarchists, *Hell's Angels: The Strange and Terrible Saga of the Outlaw Motorcycle Gangs.*

He was constantly on a soapbox about the death of the American dream or his disgust with Richard Nixon. Or was busy running on an antiestablishment platform for sheriff of Pitkin County, Colorado, which is how he ended up in *Rolling Stone*'s offices on a Saturday morning wearing dirty combat pants, a woman's wig worn sideways, a six-pack of beer in one hand, and unorganized pages from a tattered manuscript in the other.

His first article for the mostly music-driven magazine was a first-person account of his political experience. His second was "The Kentucky Derby Is Decadent and Depraved," which critics cited as his first foray into his famous "Gonzo" journalism. Hunter would later be referred to as the DNA of *Rolling Stone* magazine.

His biggest and best-known piece for *Rolling Stone* by far was "Fear and Loathing in Las Vegas: A Savage Journey to the Heart of the American Dream," which described his free-wheeling, out-of-control, acid-laced road trip to cover the Mint 400 desert motorcycle race for *Sports Illustrated*. It became a two-part series in the magazine and, later, a novel, which soon tossed Hunter into full cult status. Twenty-five years later, his career was reincarnated when Johnny Depp portrayed him in the film version.

Friends and family thought Hunter's second marriage to his assistant, Anita Bejmuk, in 2003, would be the start of a new chapter in his life. But that wasn't to be the case. That same year he had back surgery, which was followed by his accidentally breaking his leg. On a bad diet of painkillers and alcohol, he turned down jobs, failed to meet deadlines, and became odder and more reclusive. He talked to friends and family about killing himself, and obsessively polished his forty-five-caliber pistol.

Hunter could be as explosive as his massive gun collection, and as excessive and bizarre as the characters in his work. He was oddly charming, and charmingly odd. He believed that rules didn't apply to him, and for the most part, he was right. A chameleon of sorts, he was often known as Raoul Duke or "The Doctor." Other times he went by Jefferson Rank, Gene Skinner, or Sebastian Owl. But for a man who incorporated many personas and lives, the one he was no longer interested in living was his own. And as his note to Anita read, and others who knew him best always thought, thanks to his hardcore lifestyle and unhealthy behaviors, he'd lived longer than anyone expected. Sixty-seven wasn't old, but it was old enough for Hunter.

As if his massive memorial wasn't enough, on the one-year anniversary of his death, a shrine near the Gunner's View run at Aspen's Snowmass Village ski resort was created by his friends and fans. Items included numerous photos, an American flag, a gloved arm with "Gonzo" written on it, a lizard covered with multicolored jewels, an air horn, a *Rolling Stone* cover, newspaper articles, Tibetan prayer flags, and a bottle of Mr. Bubbles. Each year, the shrine continues to be updated on February 20.

UNEARTHED: Hunter's suicide closely mirrored Hemingway's. The method, the placement of the shot, the cabin in the woods—all were reminiscent of the way Hemingway killed himself. Even his feelings that he was no longer as creative as before were reminiscent of an elderly Hemingway's statement that he was "no longer the champ." And both were alcoholics. A longtime fan of the great literary figure, Hunter set out in 1964 to find out why he had killed himself. On assignment for *The National Observer*, he spent two days in Ketchum, Idaho, stopping off at Hemingway's grave, interviewing his friends and family, and making a final stop at his hero's home. The revelation was a small one: Hemingway was merely old and sick and troubled. While snooping through the dead writer's alpine chalet, Hunter pinched a high-end souvenir: an elegant pair of elk horns, which he hung proudly over the entrance to his home.

```
    City                          : PITKIN COUNTY
    Occurred after                : 17:30:00 02/20/2005
    Occurred before               : 18:00:00 02/20/2005
    When reported                 : 17:52:32 02/20/2005
    Date disposition declared     : 02/28/2005
    Incident number               : 05P000196
    Primary incident number       :
    Incident nature               : DEATH NON-CRIM  NON CRIMINAL DEATH
    Incident address              : 1278 WOODY CREEK RD
    State abbreviation            :
    ZIP Code                      :
    Contact or caller             :
    Complainant name number       :
    Area location code            : PCO4
    Received by                   : LOUTHIS, TRICIA
    How received                  : O Officer Report
    Agency code                   : PC   PITKIN COUNTY SHERIFFS OFFICE
    Responsible officer           : RYAN, RONALD E
    Offense as Taken              :
    Offense as Observed           :
    Disposition                   : ICL Closed Case
    Misc. number                  : as71/pr171
    Geobase address ID            :
    Long-term call ID             : PC0501849
    Clearance Code                : CN  CLOSED-NON CRIMINAL
    Judicial Status               : ABR  Approved by Records
= = = = = = = = = = = = = = = = = = = = = = = = = = = = = = = = = = = = = = = = = = =

INVOLVEMENTS:

Px    Record #    Date     Description                      Relationship
NM        4472  02/21/05   THOMPSON, HUNTER STOCKTON        deceased
NM       61524  02/21/05   THOMPSON, ANITA HELENKA          subject
NM       68766  02/21/05   THOMPSON, JUAN                   subject
NM       68767  02/21/05   THOMPSON, JENNIER KAY WINKEL     subject
NM       68768  02/21/05   ███████████████                  subject
CA  PC0501849   02/21/05   <Not on file>                    *Initiating Call

CASE ACTIVITY RECORD:


Seq acti activity code          when completed      ho mi officer
1    RECD Records Release       13:55:32 03/01/2005  0  0 RAAB, PATTY

LAW CASE MANAGEMENT:


Assn. Date Due Date    Status Dat Date Detai Max Sol Detl Sta Next Agen
**/**/****  **/**/****  **/**/****  **/**/****  0    0

LAW INCIDENT CIRCUMSTANCES:


Se Circu Circumstance code            Miscellaneous
1   ALET  Lethal Force
2   LT20  Residence/Home
3   BM88  No Bias
```

Hunter's police response report and autopsy

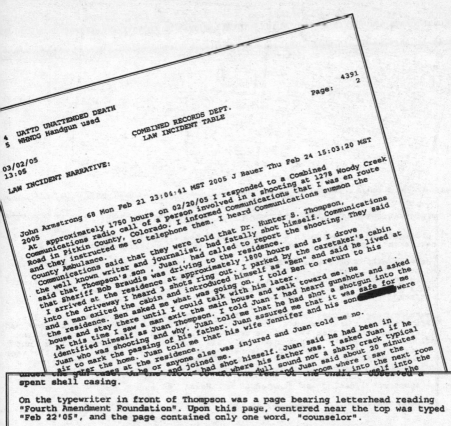

4 UATTD UNATTENDED DEATH
5 WHNDG Handgun used

COMBINED RECORDS DEPT.
LAW INCIDENT TABLE

03/02/05
13:05

LAW INCIDENT NARRATIVE:

John Armstrong 68 Mon Feb 21 23:04:41 MST 2005 J Bauer Thu Feb 24 15:03:20 MST 2005

At approximately 1750 hours on 02/20/05 I responded to a Combined Communications radio call of a person involved in a shooting at 1278 Woody Creek Road in Pitkin County, Colorado. I informed Communications that I was en route and they instructed me to telephone them. I heard Communications summon the County Ambulance.

Communications said that they were told that Dr. Hunter S. Thompson, the well known writer and journalist, had fatally shot himself. Communications said that Thompson's son , Juan , had called to report the shooting. They said that Sheriff Bob Braudis was driving to the residence.

I arrived at the residence at approximately 1800 hours and as I drove into the driveway I heard 3 shots ring out. I parked by the caretaker's cabin and a man exited the cabin and introduced himself as "Ben" and said he lived at the residence. Ben asked me what was going on. I told Ben to return to his house and stay there until I could talk with him later.

At this time I saw a man exit the main house and walk toward me. He identified himself as Juan Thompson. I told Juan I had heard gunshots and asked Juan who was shooting and why. Juan told me that he had shot a shotgun into the air to mark the passing of his father. Juan assured me that it was safe for me to be or anyone else at the residence.

...had shot himself. Juan said he had been in the room where his father was. I asked Juan if he heard where the dull sound not a sharp crack typical ...Juan said about 20 minutes ...into the next room ...self into the

spent shell casing.

On the typewriter in front of Thompson was a page bearing letterhead reading "Fourth Amendment Foundation". Upon this page, centered near the top was typed "Feb 22'05", and the page contained only one word, "counselor".

I photographed the scene in the condition I had found it, and I made measurements to Thompson and the items around him via triangulation.

the ambulance
scene.

I saw Dr. Thompson sitting upright and lifeless. There Dr Thompson was pale. There was a substance on Dr. Thompson was still and lifeless. There was a substance on Dr. ...on his chest. There was alot of blood. I lifted Dr. Thompson's oropharynx mouth. Dr. Thompson was pale. His color was pale. I saw blood on the back of Dr. to evidence breathing. His hand had a grey substance and found none. Thompson on the counter and stove. I saw blood on the back of Dr. for pulse and noticed that his hand had a grey substance and found none. pencil lead. I detected no pulse. I checked for pulse and found none. head. Both Deputy Gibson and I thought his father had used a .357 Juan Thompson told me that he thought his father had used a .357 handgun to kill himself.

Dr Thompson was sitting upright in the chair but his head was faced down, chin on his chest.

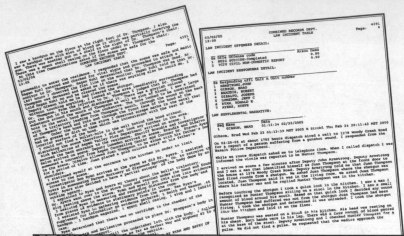

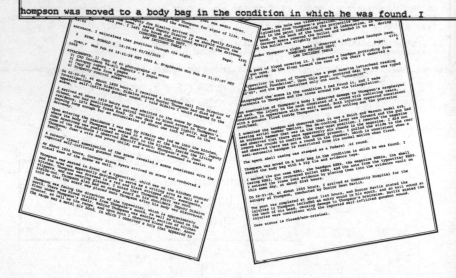

DID YOU KNOW: Sixty-four percent of men and 40 percent of women make gunshots their death of choice. Most shots are angled slightly upward, toward the desired spot. Since suicides point the gun directly on the skin, a burn mark around the wound and the presence of gunpowder residue on the hand is usually a telltale sign of a suicide as opposed to a homicide. If the gun is fired just above a bone, such as the skull or the sternum, a starlike wound is produced.

UNDERSTANDING HIS ONE-WORD NOTE: For a man who had never been short on words, for Hunter's last to have been only *counselor* left everyone confused. Many said he was trying to mimic *Citizen Kane*'s "Rosebud." Others speculated that he used the Fourth Amendment Foundation stationery to make a statement about civil rights. Since he was raised as a Christian and studied the Bible, still others pointed to the Gospel of John 14:16–17: "And I will pray the Father, and he will give you another Counselor, to be with you for ever, even the Spirit of truth, whom the world cannot receive, because it neither sees him nor knows him; you know him, for he dwells with you, and will be in you."

CAREER HIGHLIGHTS: Hunter kept carbon copies of everything he wrote. In his early years, he used to retype the works of Fitzgerald and Hemingway in order to understand their style and rhythm. In a nod to his predecessors, he would write on an old IBM Selectric typewriter. He often began writing a story in the middle, skipping around, writing specific moments he knew he wanted to include—some that were real, others that were not—knowing they would be inserted later. A weather description would generally start a piece, and last in his creative process was the conclusion, which he would call "The Wisdom." He rarely met a deadline, his notes were hard to decipher, and since some parts of what he wrote were made up, fact-checking his work was close to impossible. If he was not personally supervised by an editor or publisher, it was doubtful he'd complete an assignment. His greatest achievement was creating "Gonzo journalism," a first-person, in-your-face style of writing that blended fact and fiction together seamlessly.

At the end of his life, his career came full circle when he wrote about sports. His five-year weekly column, "Hey Rube" for ESPN. com's "Page 2," resulted in the collection *Hey Rube: Blood Sport, the Bush Doctrine, and the Downward Spiral of Dumbness—Modern History from the Sports Desk*.

A dozen documentaries on Hunter Thompson have popped up over the past decade, most recently *When I Die*, which chronicles the making of his farewell wishes into a reality. And Johnny Depp will be channeling Thompson once again when he plays Paul Kemp in an adaptation of Thompson's *The Rum Diary*. Originally written in 1962, *The Rum Diary* fictionalizes Thompson's experiences in Puerto Rico. The book took more than forty years to be published.

Sylvia

Sylvia Plath

BORN: October 27, 1932, Jamaica Plain, Massachusetts
DIED: February 11, 1963, Court Green House, Devon, England
AGE: 30
METHOD: Asphyxiation
DISCOVERED BY: The nanny
FUNERAL: Among the long list of eminent writers present at her funeral, close friend Anne Sexton gave a touching eulogy and talked openly about the two women's attraction to suicide.
FINAL RESTING PLACE: Heptonstall, Yorkshire

Dying
Is an art, like everything else.
I do it exceptionally well.
I do it so it feels like hell.
I do it so it feels real.
I guess you could say I've a call.

FROM "LADY LAZARUS," IN *THE COLLECTED POEMS*, 1962

The first time Sylvia Plath tried to kill herself, she was almost triumphant. A driven, focused overachiever, the twenty-year-old poet prodigy perfectionist rarely failed at anything. Plain and WASPy looking, she often wore her shoulder-length dirty blond hair pulled back or neatly combed. Her skin was pale, her lips full, and her eyes seemed intent on soaking up every detail in a room. Her voice punched and overarticulated as witty and sharp observations popped from her mouth.

It was the passing of her father that set her depression and lust for death in motion. Nine days after her eighth birthday, he died from diabetes mellitus, leaving her feeling abandoned, punished, and deliberately betrayed.

For a writer who flourished and feared during the Great Depression and World War II, which immensely affected her mental and emotional status, Sylvia began the early 1950s amazingly positively. She started publishing her poetry, writing articles for the local paper, and selling her short stories, one of which won her a guest editorship at *Mademoiselle* magazine. She'd already been honored with a scholarship to Smith College and had completed her freshman year when, in the summer of 1953, she returned home disillusioned and frustrated by her negative experience with the publishing industry. She was mentally, emotionally, and physically exhausted. She thought her spirits would lift if she were able to study at Harvard's prestigious summer writing program, but she was turned down. Fastidious with her journal writing, something she had done since the age of twelve, she nonetheless divulged hardly any important information in her diaries during this period, and they end abruptly in July.

By August, she was in a deep depression. She couldn't eat, focus, or write. She combed bookstores and libraries, hungry for psychiatric books, all of which confirmed that she was losing her mind. She made slit marks on her thighs in an attempt to release her pain, and when her mother saw them, Sylvia broke down, insisting that she wanted to die, that they could both die together.

Days later, while picnicking on a double date at the beach, she ran into the ocean, trying to get sucked up by the tide. She emerged a wet failure. She needed a better plan.

She found one on August 24, 1953. At 2:00 PM, she waited for her mother, who was going to the movies, to leave the house. Finally free, Sylvia broke into the metal box that held a bottle of sleeping pills, took a blanket and a jar of water, and proceeded to the dining room, where she left a note on the table: "Have gone for a long walk. Will be home tomorrow."

She went down to the cellar, shoved a stack of firewood away from a two-and-a-half-foot crawl space underneath the family's screened-in porch, where her grandparents sat sunning themselves above her, and climbed inside. The bottle she'd taken contained forty-eight pills. One by one, she placed them on her tongue, washed them down, and then cocooned herself like a caterpillar in the blanket, waiting for the pills to take effect.

Miles away, in a dark movie theater, her mother suddenly sensed that something was wrong. Against her better judgment, though, she stayed till the end of the film, and returned home later to find Sylvia's note.

By morning, stories about a "Smith girl who had gone missing" filled the newspapers. Radio stations also piped out information about Sylvia. More than one hundred people looked for her. Search parties scoured the lakes, woods, and beaches. Friends checked her favorite Boston haunts. Police roamed the streets. After missing for more than forty-eight hours, she was found accidentally when her grandmother, who felt she needed to resume her daily activities and do the laundry in the basement, heard a moaning coming from the crawl space.

When they pulled her out, she was barely conscious; her face was bloody and covered in dried vomit. A gash from hitting her head on the concrete upon waking was already infected. Eight pills were found in the bottle, indicating that she'd swallowed forty, but she'd thrown them up before they were able to take effect.

After a short stint at the hospital, she was sent to the famous McLean Hospital to recuperate and receive electric shock treatment. Able to pull herself together, she was readmitted to Smith, where she graduated summa cum laude in 1955, then won a Fulbright, which allowed her to study for a year at Newnham College at Cambridge University. Though the accolades were esteem builders for her, she would always see herself as the girl whose father had left her. The girl who tried to kill herself. The girl whose family was poor.

Sylvia met Ted Hughes in 1956 at a party in Cambridge thrown by a group of writers who'd started a literary magazine. Though he had brought a date, the two began talking. There was drinking, dancing, intense conversation, and before she left, she bit him on the cheek, leaving a mark that lasted for days. They corresponded through their poetry, and were married four months later. It took only a few months for Ted to have an affair, with Assia Wevill, a married poet who was renting a house from the Hugheses.

By 1960 Sylvia and Ted each had dueling books of poetry published, and she had given birth to their first child. Over the next few years, she far surpassed his accomplishments. In 1961 she wrote twenty-two poems and won a $2,000 Saxton Grant to work on a novel (which was already finished). The following year she penned twenty-five more poems, all within a month, and had their second child, Nicholas. Yet Ted's infidelity with Assia greatly contributed to Sylvia's unhappiness. The marriage fell apart. And in late 1962, Ted left her and their two children. Not one to accept defeat, Sylvia rented a house at 23 Fitzroy Road, London. Though she moved there with the naïveté of a child, drunk on the fact that W. B. Yeats had once lived there, she didn't count on many things: the brutal, bleak, and unrelenting winter, which would catapult her deeper into depression; the nanny quitting, leaving her to care for the children on her own; the pipes freezing; the lack of heat or a telephone. Ice and snow blanketed the streets. The bad weather increased the isolation she already felt. Her experience epitomized the expressions "cabin fever" and "going stir crazy."

In the days leading up to her death, her doctor grew frighteningly alarmed. He sent a letter with the names of several psychiatrists, which was lost in the mail and never delivered. He tried to secure a hospital room for her, but failed at that as well. Overwhelmed by the very act of breathing, Sylvia went to stay with friends for a weekend, comforted by knowing that a nurse would be there to greet her Monday morning, as her doctor had arranged.

On Sunday evening, she insisted upon going home, rather than staying until Monday morning. At 11:00 PM, she found herself knocking on her neighbor's door, desperate for postage stamps. Though the neighbor said she could have the stamps, she insisted on paying for them, then asked him what time he left for work in the morning. He later reported that she seemed groggy; he assumed she was ill.

Back in her apartment she paced incessantly and wrote several notes, one which she placed near the front door. Printed on it was her doctor's name and phone number. It simply asked someone to phone him. She placed mugs of milk and some bread and butter by her children's bed while they slept upstairs. After shutting their door, she sealed their room with masking tape. *Ariel and Other Poems*, a newly completed manuscript, lay on the table near the front door, like a present waiting to be opened.

Downstairs she worked methodically, placing tea-soaked cloths into holes and vents. In the kitchen she sealed the door, and as if finishing the botched job she began twenty years before, she completed the task. The oven was a cheaper method than pills, and a folded towel acted as a substitute for a blanket, which she used to support her head on the stove's open door. The kitchen was a larger cave than the small crawl space she tried to use years before, but it filled quickly with the gas. By 6:00 AM, one of the greatest poets of the twentieth century was dead at the age of thirty.

As her doctor had assured her, a nurse arrived at 9:00 AM ready for work. The door was locked and Sylvia's name wasn't written on the ringer. Confused, the nurse rang Sylvia's neighbor and, after receiving no answer, walked to the coin box to phone her employment agency; she was told to wait at the house. Upon returning, she saw Sylvia's children crying in the upstairs window. Panicking, she ran for help. She found a construction worker, who broke down the door. The toxic smell of gas hit them immediately. They forced their way into the kitchen and found Sylvia with her head still resting on the oven door. While the workman carried Sylvia's lifeless body into the living room, the nurse turned off the gas and opened the windows. Then she started CPR on her. After finding the note Sylvia had left instructing someone to phone her doctor, they summoned Dr. Horder, who pronounced her dead at 10:30 AM.

Many critics and specialists say Sylvia's methodical planning of her suicide—leaving the note, timing the nurse's arrival, determining when her neighbor would leave for work—all indicated that she wanted to be saved, that she was secretly hoping history would repeat itself. Many also feel that the poem "Edge," which she wrote six days before her death, was her version of a suicide note, describing what she hoped killing herself would accomplish.

Just as she didn't count on vomiting up the pills, she also didn't realize that her neighbor's bed was directly under her stove, and that the gas would leak through the holes and cracks in her kitchen and into his bedroom. When they found her neighbor, he was unconscious. Though they successfully revived him, the hospital said he had carbon monoxide poisoning.

Her family found out about her death through Ted, who sent a callous note to her aunt, stating, "Sylvia died yesterday."

On February 15, friends and family piled into St. Pancras County Court, a claustrophobically damp room where Sylvia's "sudden death" inquisition took place. Peter Sutton reported on the postmortem; Ted identified the body. He then traveled with it to Yorkshire, where it was prepared for burial. The funeral was held the following day at a local church. Small and intimate, the service was conducted by a minister who barely knew the couple. The mourners went to the cemetery afterward to continue eulogizing Sylvia. Her mother didn't attend; neither did her two children, which only fed into the terrible cycle created by Sylvia's mother when she refused to let Sylvia attend her father's memorial, a decision Sylvia never forgave her for.

The inscription on her headstone, along with a line from the Bhagavad Gita—"Even amidst fierce flames the golden lotus can be planted"—was chosen by Ted.

Shortly after Sylvia's funeral, her friend Elizabeth was sent a letter by Assia, now Ted's wife. It contained the gas bill, which covered the period of February. "She was your friend. You pay the bill" was all it read.

UNEARTHED: For years, suicide was made easy for many Britons, since homes were now heated and ovens—often called "execution chambers"—were fueled by coal gas. Plentiful and cheap, in its unburned form, coal released ultra-high levels of carbon monoxide. One could easily get asphyxiated thanks to an open valve or a leak in a closed space. Cheaper than pills, easier than hanging, and less messy than the remnants from a gunshot, death by coal was quick, painless, and, most important, easily accessible. By the late 1950s it accounted for some 2,500 suicides a year in Britain, almost half the nation's total. This raised a red flag, and led to a new program to substitute natural gas for coal. By the early 1970s, carbon monoxide-fueled homes and ovens were almost nonexistent. Though the nation's coal-induced suicide rate dropped significantly, that of other methods rose over the same period.

TWO WIVES, SAME METHOD: Six years after Sylvia's passing, Ted's second wife, Assia, killed herself and their four-year-old daughter in a pathologically eerie way.

In March 1969, the couple had had an argument over the phone. Though Assia had set the table for lunch and was expecting Ted home, she sent the nanny on an errand, dragged a mattress into the kitchen, sealed the door and window, and turned on the gas. She and her daughter drank sleeping pills dissolved in a glass of water, then lay down and went to sleep under the woozy, fuzzy fumes from the oven.

When the au pair returned, the two were dead.

Assia and Ted, who was her fourth husband, had been together for eight years; the last four were rocky, with him away most of the time, leaving Assia lonely and unsure of his faithfulness. Though she also wrote poetry, after Sylvia's death she'd been snubbed by the writing world and by Ted's family and friends.

A suicide note written to her sister was found: "I am clearly ill. The last four years have been a strain simply too hard to bear." She had packed two trunks' worth of her daughter's belongings, wrote a check for $1,200 to cover the fare to Canada, where her sister lived, and enclosed poems she'd stolen from Sylvia—in case she ever needed money or something important to barter with. But since Shura was found dead alongside her mother, Assia's suicide was thought to be an impulsive act, reflective of a woman who'd suddenly snapped.

ANOTHER DEATH: In 2009 Sylvia's forty-seven-year-old son, Nicholas, was found by his girlfriend having hanged himself in his Alaskan home. Only a year old when his mother died, the professor of fisheries and ocean studies battled depression for much of his life. As if trying to break a suicide cycle, neither he nor his sister chose to have children. His father, who died from cancer in October 1998, was fortunate not to be a witness to yet another family tragedy.

CAREER HIGHLIGHTS: Sylvia's strange, early death further orbited her into cult status, making her feminism royalty. Much of her feelings of anger and abandonment surrounding her father bleed into her poetry, such as "Daddy" and "The Colossus," just as her stay at McLean and her stint at *Mademoiselle* were the impetus surrounding *The Bell Jar*, which was published a month before her death. Though she wrote the autobiographical novel under the pseudonym Victoria Lucas, her real name was later used, and the book became a reading rite of passage for the under-thirty crowd.

Many of the frighteningly accurate diaries Sylvia kept throughout her life were published after her death. *The Journals of Sylvia Plath* earned her a posthumous Pulitzer Prize in 1982. One of the biggest losses is the journal she kept after *The Bell Jar* and up until three days prior to her death, which Ted burned, stating that he didn't want people to see Sylvia's downward spiral. Some critics say he was afraid his already tarnished persona would be condemned further by her remarks about him. However, he did make available for publication *Ariel and Other Poems*, the manuscript Sylvia left on the table the night she died. Today, two of her journals are on exhibition at Smith College, where they will remain until 2013, the year marking the fiftieth anniversary of her death.

Anne Sexton sharing her book of poems *Live or Die*, May 1967

Anne Sexton

BORN: November 9, 1928, Newton, Massachusetts
DIED: October 4, 1974, Weston, Massachusetts
AGE: 45
METHOD: Carbon monoxide poisoning in the garage of her home
DISCOVERED BY: Her house-sitters

FUNERAL: At Anne's memorial service, poet Adrienne Rich described the self-indulgence of suicidal personalities; poet Denise Levertov noted in an obituary that Anne had confused creativity with self-annihilation. Music was performed by her poetry-rock group, Her Kind.

FINAL RESTING PLACE: She was buried in a plain pine box—which she had requested—at Forest Hills Cemetery and Crematory in Jamaica Plain, Massachusetts. On her headstone she wanted a palindrome—"Rats Live on No Evil Star"—stating that this metaphor, used to describe her sick self, gave her a peculiar feeling of hope.

> *Sylvia Plath and I would talk at length about our first suicide, in detail and in depth. . . . Suicide is, after all, the opposite of the poem. Sylvia and I often talked opposites. We talked death with burned-up intensity, both of us drawn to it like moths to an electric light bulb.*
>
> FROM "THE ART OF POETRY NO. 15"
> (interview in *The Paris Review*, Summer 1971)

Standing in her kitchen, she pours herself a generous glass of vodka, her lubricant of choice, and dials up her date, hoping to change the time of their meeting. She removes each ring from her fingers, stashing them in her purse, and retrieves her mother's oversize fur from her closet. She downs her drink and retraces her steps, generously refills her glass, and proceeds to her garage. Her red 1967 Cougar is a beautiful sight. At this moment, Anne Sexton is surrounded by the items most important to her: her car, her vodka, her mother's fur. She climbs into the driver's seat, revs up the engine, turns on the radio, and waits. The slight burning sensation of cool liquid slipping down her throat is a distraction from the effects of the exhaust fumes. She has waited a lifetime for this moment.

A love of all things bad—liquor, pills, cigarettes, promiscuity—had Anne on a slow, suicidal path for decades. She wrote about suicide, confessed to wanting it, was obsessed with it. She was hospitalized twenty-two times, with more than nine attempts to her credit. It's amazing the edgy, biting, and provocative poet didn't die sooner.

Anne was methodical, planning her demise for more than a year, during which time she put her house in order: asked friends what possessions of hers they wanted, selected a biographer, hired a secretary to help prepare an archive of her work, and drew up a will with specific instructions for her funeral. She even waited until her daughter, Linda, was old enough to be her literary executor.

Five years earlier she'd written a note to Linda while on a flight to St. Louis to give a reading, recounting how much she loved her, how terrible it was to be forty, motherless, and how sorry she was. "Life is not easy. It is awfully lonely. *I* know that," she wrote in April 1969. "Talk to my poems, and talk to your heart—I'm in both; if you need me. I lied, Linda. I did love my mother and she loved me. She never held me but I miss her, so that I have to deny I ever loved her—or she me!" She signed the letter "XOXOXO Mom."

During her final thirty hours, she left a trail of activities. On October 3 she returned home to Boston, having given a paid reading at Goucher College in Baltimore. As a special treat, she let her Boston University poetry class meet her at the airport. The next morning she stayed in bed until a friend arrived. The two had breakfast, and then Anne drove to Cambridge for a therapy appointment. The date of October 4 hadn't been picked arbitrarily. Anne was methodical and clever, premeditative and calculating. She'd been a patient of Barbara Schwartz's for exactly nine months to the date of this final session. The lack of love she felt from her mother dominated many of the issues she discussed with Schwartz, and as any Freudian thera pist will note, Anne felt as though Schwartz had given birth to her. In one of her last poems, "The Green Room," which she had written for Schwartz in gratitude for the mothering she provided, she called the therapist a "consecrating mother." After that last session, Anne deliberately left her cigarettes and lighter behind in the doctor's bowl of daisies.

The next stop was lunch with good friend and poet Maxine Kumin. The goal was to proofread galleys of her new book of poems, *The Awful Rowing Toward God*, which would be published in March. By 1:30 PM, they were done, and Anne returned home, poured her drink, ice clinking seductively in the glass, put on the cherished fur, got into the car, sealed the door, and at 3:00 PM on that sunny afternoon, died.

A college dropout turned housewife, fashion model, librarian, sales girl, and jazz singer, Anne Gray Harvey was the youngest of three sisters born to a difficult, alcoholic father and a callous, malicious mother. Torturing young Anne by telling her she wasn't the favorite child, her mother blamed her for the cancer she died from and didn't believe her daughter when she showed her poems she said she'd written. In August 1948, at age nineteen, she eloped with Alfred Sexton, nicknamed "Kayo" after a cartoon character, whom she had dated for only a month. The duo drove to North Carolina, where the legal marrying age was eighteen.

The strong-willed Anne was tall, possessed a powerful physical presence, and was thought of as a seductive Anne Bancroft–like glamour girl, with wavy black hair, penetrating blue eyes, and long, shapely legs. Her voice was smoky, her lips always painted with red lipstick, and she chain-smoked Salem cigarettes. She once realized her purse was so heavy because it contained fifty-five Bic lighters and bottles of pills: antidepressants, tranquilizers, and sleeping agents.

Anne had a terrible childhood, and claimed she was locked in her room until the age of five. Unable to find nurturers in either parent or her sibling, she turned for attention to her grandmother, her "twin," as she called her, and the two developed an unhealthy relationship. Over the years, this became her modus operandi: blurring the lines and boundaries with most people in her life, especially with her two daughters, who, when no one else was available, became her sisters, her friends, her caretakers, and, in Linda's case, her lover.

Boston in the mid-1950s was bursting with poets, and she was another writer who flourished in postwar America. Her themes of madness, sexuality, death, abortion, incest, masturbation, and adultery made her popular among the feminists. Like Virginia, like Sylvia, Anne was brilliant, ahead of her time, and tragically unhappy. After giving birth to her daughter Linda in 1953, she fell into a deep postpartum depression, which was revisited with her second daughter, Joy, two years later.

Anne seemed almost attracted to any disease that sounded interesting, often claiming to pick up new illnesses from the many hospital stays she required. A depressive alcoholic addicted to tranquilizers, she was also diagnosed with hysteria, agoraphobia, anorexia, insomnia, severe mood swings, and bipolar disorder. Undiscriminating when it came to sexual partners, she had affairs with at least one of her therapists and many other random men. She claimed she heard voices, thought she'd been molested, and developed an alternate persona named Elizabeth. When frustrated, she would throw tantrums, stamping and screaming like a three-year-old. She was selfish and narcissistic. Due to her inability to care for her own children, others often had to step in to help.

In 1956, on her birthday, and with Kayo away on a business trip, she made her first suicide attempt by overdosing on Nembutal—which she called her "kill-me pills." After a stint in a mental hospital, she began seeing Dr. Orne. During her first session, she told him her only real talent might be prostitution. He suggested she write instead. After she watched a half-hour television show titled "How to Write a Sonnet," a pen was never out of reach.

A second suicide attempt happened a year and a half later, after another hospital appearance. Still, she continued to write, and in 1958, she enrolled in Robert Lowell's graduate writing seminar at Boston University. There she met poets Sylvia and George Starbuck, among other literary hotshots.

In 1959, her father died from a stroke, and her mother, from cancer a few months later. "All My Pretty Ones" grew out of her grief and guilt over their deaths. Unable to function, to parent, to keep a home or husband, or to live sober, she nonetheless somehow managed to write. She won awards and published—her work accepted by *The New Yorker* and *Harper's* magazine. Meanwhile, she watched her marriage fall apart. And so it went. Hospital stays, suicide attempts, failed relationships. Publication.

In 1973 she was hospitalized three times and received a divorce from her husband. A year before she died, she tried overdosing again, using a mixture from a number of pills she had acquired. Found comatose a day later, she was rushed to the hospital, where the initial diagnosis was severe brain damage. Surprisingly, she emerged from her coma without repercussions. Her last hospital stay was at McLean, which she left looking disheveled and thin. She would live only another eleven months.

"The pills are a mother, but better," she wrote in her poem collection *The Addict*. "Every color and as good as sour balls. I'm on a diet from death."

Sadly she wasn't, and the addict traded in her mini savors, which had failed her before, and turned instead to a revving engine and a sealed garage door.

UNEARTHED: Carbon monoxide is a colorless, odorless toxic gas that interferes with the delivery of oxygen throughout the body. Because oxygen cannot attach to hemoglobin as quickly, the blood system is left with less and less of it for fuel. This yields a slow, toxic suffocation. Carbon monoxide poisoning can result in a variety of symptoms, ranging from headache, to weakness, lethargy, nausea, confusion, disorientation, seizure, and fatality. Once death occurs, the lips and the skin can become cherry red. Sometimes a blue-gray color from lack of oxygen, called cyanosis, is seen in the beds of fingernails and toenails and on the lips and tongue. Medical examiners during an autopsy might look for reddish color on the internal tissues. If the body is found several hours or a day later, the skin will be loose from the elevated temperature caused by the car's exhaust getting into the passenger compartment. Everyone has a small amount of carbon monoxide in their blood all the time. Five percent is considered a safe level, though some smokers can have up to 13 percent. In a healthy person, a level of 50 percent will cause unconsciousness.

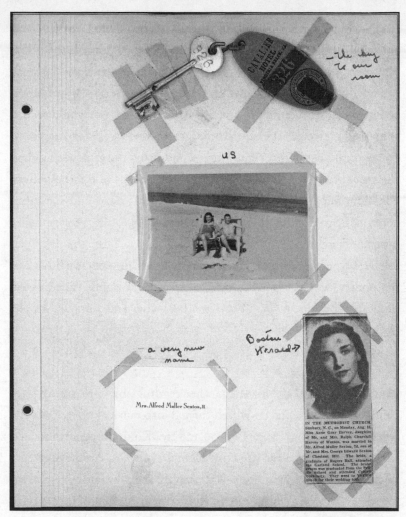

The day to our room

us

a very new name

Boston Herald →

Mrs. Alfred Muller Sexton, II.

IN THE METHODIST CHURCH, Sunbury, N. C., on Monday, Aug. 16, Miss Anne Gray Harvey, daughter of Mr. and Mrs. Ralph Churchill Harvey of Weston, was married to Mr. Alfred Muller Sexton, 2d, son of Mr. and Mrs. George Edward Sexton of Chestnut Hill. The bride, a graduate of Rogers Hall, attended the Garland School. The bridegroom was graduated from the Perkins School and attended Colgate University. They went to Virginia Beach for their wedding trip.

Anne's scrapbook, page 74

THE TAPES: Anne was controversial throughout her entire life, so it made perfect sense that controversy would follow her in death. When *Anne Sexton: A Biography*, by Diane Wood Middlebrook, was finally published, it had already been through much scrutiny over the biographer's use of recordings made during Anne's therapy sessions, which her psychiatrist agreed could be included in the book. The three hundred hours' worth of tapes had been made because Anne would go into trances during her therapy sessions and, upon waking, have no memory of what took place. Dr. Orne would record the session, give her the tape from the previous week, and ask her to go home and listen to it, make notes, and bring the tape back. When the tapes were later used in the biography, issues of doctor-patient confidentiality arose, especially since she was dead and the tapes contained discussions of how she had had an incestuous relationship with her daughter.

Dr. Orne argued that Anne, who held nothing personal back in her poetry, would have wanted her tapes made public. The Sexton biography became a *New York Times* bestseller and a finalist for the National Book Award.

CAREER HIGHLIGHTS: Anne began her readings by performing "Her Kind," a signature piece—which turned into musical poetry à la the chamber rock group Anne Sexton and Her Kind. Along with receiving a Pulitzer Prize for the collection *Live or Die*, Anne received a number of kudos, including a National Book Award nomination for *All My Pretty Ones*, an American Academy of Arts and Letters traveling fellowship, the Shelley Memorial Prize for *Live or Die*, and a Guggenheim Foundation grant. *Mercy Street*, her two-act play, was performed at the American Place Theatre in 1969–70. Toward the end of her life, her poetry developed more religious themes, which dominated her final collection, *The Book of Folly*. *The Awful Rowing Toward God* was the first volume of poetry to appear after her death.

45 Mercy Street and *Words for Dr. Y* followed. Her poems, always edgy and self-revealing, were riddled with themes of death, addiction, and suicide, most noticeably those in *The Addict, Live or Die,* and *The Death Notebooks.*

Authors Unearthed

FATHER FIGURE: Of the five authors highlighted in this book, each dealt with abandonment issues. Hemingway's father shot himself when the author was twenty-nine. Hunter's died when he was fourteen. Virginia was twenty-two when her father passed away, which caused her second nervous breakdown. Sylvia's died days before her eighth birthday. And Anne, who lost both parents during the year she turned thirty-one, never felt loved or wanted by hers. Worst was John Berryman, who was eight when his father walked outside their home and shot himself below his son's bedroom window. For all these fatherless figures, the profound loss greatly affected their work, and for many, it became the thematic thread of their material.

LYSOL POISONING: As far back as one can trace, poisons have always been methods of suicide. Two of the most bizarre suicides occurred during the mid-1900s, by poets who opted for death by Lysol. (The disinfectant was used in the 1920s as a vaginal douching product, suggesting it prevented infection and odor. It was also peddled as birth control before being marketed in its present role, as a household cleaner.) In 1928, Charlotte Mew, whose work encompassed both Victorian poetry and modernism, was depressed after her sister's death. She was admitted to a nursing home, where she killed herself by drinking the lethal liquid. Vachel Lindsay, Sara Teasdale's lover, followed suit three years later, but penned a note ("They tried to get me—I got them first!") before ingesting the concoction.

Lysol deaths got the attention of playwright Tennessee Williams, who wrote two plays in which characters choose Lysol to end their lives. *Suitable Entrances to Springfield or Heaven* was a specific nod to Lindsay's suicide.

ANOTHER NOTABLE: A writer who helped define the new millennium was David Foster Wallace, whose fans came to admire the long-haired, do-rag-sporting, tobacco-chewing writer. In September of 2008, his wife found him hanging in their California home. At age forty-six, the *Infinite Jest* novelist was considered by colleagues to be a brilliant writer—challenging, warm, sincere, the best of his generation. The author of nine books of fiction and nonfiction, he'd suffered from depression for years, and when his medication's side effects became too much for him, he stopped taking his pills and opted for electro-shock therapy, which also failed him.

Four

Actors

Actors are anything but a rare breed. They're a dime a dozen, in fact. For every actor who makes it, hundreds don't. For every performer who drops out of the fame and stardom race, thousands are willing to take his place. And yet, out of all the groups in this book, actors are perhaps most accessible to us. With actors on a successful sitcom, we can spend years visiting with them weekly in thirty-minute episodic intervals. Film favorites appear larger than life and up close and personal at the same time. We want to dress like them, know where and what they eat, whom they're dating, when they get married, have children. We put them on pedestals, glorifying them into godlike figures.

Our willingness to feed their need for attention strokes their egos—which many suffer from too much of. Paired with a love for the spotlight is a gnawing insecurity, which many can't confront—often they can't perform at all—without the help from illegal drugs or booze from a bottle.

"Alcoholism causes egomania, which impels many addicts to overachieve. This is the best explanation for the fact that roughly 30 percent of Academy Award-winning actors have been alcohol or other drug addicts," says Doug Thorburn, author of *Alcoholism Myths and Realities*. "Since addicts are far more likely to commit suicide than others, celebrities are seen as committing suicide at higher rates than the general population."

For these very public people, their acts of suicide are usually performed in private, done without the paparazzi, TV crews, set designers, or fans. The actors are often found alone. Unlike the musicians and writers who committed suicide while family members were in the next room, lovers in the house, or children in bed, the following actors chose solitude, away from those who knew them best.

In 1962, Auntie Em fans were sad to learn that *The Wizard of Oz* actress Clara Blandick had killed herself in her Los Angeles home. She'd organized her belongings, placed her favorite photographs and memorabilia in prominent places, and arranged her résumé and collection of press clippings before swallowing a bottle of sleeping pills. Dressed in a royal blue gown, her hair properly styled, she lay down on a couch, covered herself with a gold blanket, and tied a plastic bag over her head. She and her suicide note were found Sunday morning by her landlady.

Dave Garroway, one of Chicago's most beloved anchormen and the first host of NBC's *Today* show, shot himself in his home in Philadelphia in 1982. Though his body was discovered in the morning by his housekeeper, his wife had seen him alive just fifteen minutes earlier.

His method is not unusual. It's been copied over and over by actors. In 1990, Rusty Hamer, the actor who played the son in the 1950s show *Make Room for Daddy*, shot himself in his home. In 2006, police found Ben Hendrickson, the twenty-one-year veteran of *As the World Turns* and film roles, dead in his bedroom with a gunshot wound to his head. Before there was Mini Me, there was Hervé Villechaize, who gained worldwide recognition for his role as Mr. Roarke's assistant, Tattoo, in the popular '70s show *Fantasy Island*. (James Bond fans knew him as the evil henchman Nick Nack in *The Man with the Golden Gun*.) Depressed and drunk, erratic and violent, Villechaize shot himself in his Los Angeles home in 1993. And thirty-six-year-old Brenda Benet, a soap actress from *Days of Our Lives*, who divorced TV's *Incredible Hulk*, Bill Bixby, aimed for her mouth with a Colt .38. She was found on the floor surrounded by a circle of lit candles.

For many actors, their methods are quick, their approach determined. Hanging, jumping, gunshots—each ensuring a high success rate. For Clara Blandick, the need to ensure her death was so high that she chose the additional method of asphyxiation in case the pills didn't work.

What all these suicides have in common is their solitary nature. Actors seem to, as Greta Garbo is claimed to have uttered, "want to be alone." There is an obvious irony in the seclusion they seek. Despite their need for approval, applause, and recognition—the reasons why many of them turn to acting in the first place—in the end, overwhelmed by depression and too much attention, and filled with loneliness, they choose isolation.

Spalding performing from his monologue *Impossible Vacation* at the State College, Pennsylvania, in March 1991

Spalding Gray

BORN: June 5, 1941, Providence, Rhode Island

DIED: January 10, 2004, New York City

AGE: 62

METHOD: Drowning

DISCOVERED BY: A stranger on March 7, when Spalding's body washed up on the waterfront by Greenpoint, Brooklyn

FUNERAL: On April 13, 2004, a memorial service was held for Spalding at New York City's Lincoln Center, in the Vivian Beaumont Theater, the site of several of his performances. Attendees included Francine Prose, Philip Glass, Eric Bogosian, Eric Stoltz, and John Perry Barlow, a lyricist for the Grateful Dead.

FINAL RESTING PLACE: Spalding was cremated and buried at Oakland Cemetery in Sag Harbor. He's honored in New York City, with a plaque in Washington Square Park and a paving stone in Tompkins Square Park. He provided an epitaph for his own tombstone in a 1997 interview: "An American Original: Troubled, Inner-Directed, and Cannot Type."

I only see through loss, death and my right eye. I only start to see the color of everything, the intensity of everything when I'm leaving.

IN AN INTERVIEW FOR *IO* MAGAZINE

The night is cold—eleven degrees Fahrenheit. It's the kind of cold that bites into your skin and bulldozes its way through your bones. Standing on the outer deck of the ferry from Manhattan to Staten Island is almost masochistic. Water slaps harshly against the small boat, and the wind seems even beastlier.

Few people took the trip the night of Saturday, January 10, 2004. Many stayed home, opting to rent a movie and order in, as New Yorkers often do during wretched weather. On the ferry that evening was monologist Spalding Gray, who had finally decided to explore his intense attraction to death. He merely walked to the edge of the back deck, ignoring the thick rope that stretched across the doors—a recent addition, to deter people from jumping—stepped over the black railing, entered the wet darkness, and disappeared.

The magician waves his wand and *poof*—all gone.

As in a magic act gone incredibly wrong, Spalding disappeared—to reappear two months later, dead. He was found by Robin Snead, who was walking by the Brooklyn waterfront and noticed a pair of legs poking out from under the pier. At first he thought it was a joke or an installment by one of the many innovative artists who live in

Williamsburg. Lying in a perfect pose of crucifixion, waves overlapping against his floating body, was Spalding. There were no flies. No pungent smell. The body, which bore a plastic brace on one of the legs, was pale, dressed in black corduroy pants, a red flannel shirt, and sneakers.

In the winter, floaters sink to the bottom of the ocean, where they are dragged about, their faces rubbing up against the bottom of the sea. As the weather warms, they rise back up, reemerging into the world. When the police arrived and turned him over, they found that his face had been completely eaten away. Only his eyes remained. He could have been anyone.

Dental records later confirmed the body as Spalding Gray's.

Spalding was a minimalist monologist. Some might call his work poor man's theater. He performed his life stories with a quiet, reserved mania. A desk, a spiral notebook, a glass of water, a mic, and Spalding were usually the only things onstage. With his autobiographical style of storytelling, he paved the way for a number of monologists, especially during the 1980s, commercializing the craft and turning it into a respectable art form. He'd pull anecdotes exclusively from his life—about his travels, obsessions, fascinations, and revelations—sharing intimate and raw pieces of himself with his audiences.

Sex and Death to the Age 14 was about his childhood, mixed with world events. *Monster in a Box* revolved around his novel and the publishing process. In *Gray's Anatomy*, he explained his search for answers to a rare disease that had damaged his vision in his left eye. For *It's a Slippery Slope* he paired his passion for skiing with his role as a new father. *Swimming to Cambodia*, his most well known and acclaimed show, revealed his life-altering experience in Roland Joffé's movie *The Killing Fields*, which explored the relationship between a *New York Times* reporter and a Cambodian photographer. Over the next two years, he developed a film version of *Swimming to Cambodia* with director Jonathan Demme. The stage show won him an Obie (an Off-Broadway theater award), and the film was nominated for four Independent Spirit Awards.

The middle son of WASP parents, Spalding grew up in Rhode Island with his two brothers, Rockwell Jr. and Channing. At Emerson College he pursued acting while trying to perfect the craft he became known for. In 1977 he co-founded the Wooster Group, an experimental theater company in Lower Manhattan.

He had a long, powerful relationship with death. Like a lover, he flirted with it often. And death called to him endlessly, begging the monologist to slip down the rabbit hole and vanish. In 2004 he did.

It was no secret that he battled hereditary depression. It had killed his mother at age fifty-two, when she locked herself in her car and let the engine run in their sealed garage. He talked openly about his mental anguish over and over in his shows; it almost fueled them. It was his material, his shtick.

Plagued by claustrophobia, obsessive-compulsive thoughts, depression, anxiety, and bipolar disorder, Spalding was ironic, neurotic, fearful, and exuded a sense of satire. He paced often. He longed for many things: for his mind to quiet, for the depression to momentarily halt, to find what he called "the perfect moment," which he'd experienced only a few times, mostly while he was skiing.

In June 2001, during a vacation in Ireland celebrating the monologist's sixtieth birthday with his wife, Kathie, and a few friends, a massive car crash on a narrow road that locals call "the black spot" altered everything for him.

"We were five adults in the car which stopped to turn right on this very narrow road and this guy came around the corner in a van," Spalding told an interviewer for the *Harvard University Gazette*. "He hit us. I was in the backseat and Kathie was driving. I flew forward, impacting my head on hers. Her seat came back. The engine went right into the cabin. I think what happened was the seat pushed my femur, dislocated my hip and fractured my skull. Next thing I know I was in a puddle of blood on the road. It was an hour before the ambulance came. It changed my life, the accident. Everything was fine and then five seconds later, I was lying in a puddle of blood." Tara Newman, a real estate broker and friend who was sitting next to him, took one look at Spalding and was sure he was dead.

He spent a month in two different hospitals before being well enough to travel home. He had a badly broken hip, which almost immobilized his right leg. A fracture in his skull had caused damage in his right frontal lobe, which went undetected for a long time. And he had a huge, jagged scar on his forehead. Over the next two years he entered twelve hospitals—two of which were mental wards—and had six operations, including the insertion of a metal plate into his head to fix the fracture, which left a concave mark in his face. He was put on a cocktail of antidepressants, antipsychotics, and anti-seizure medication, and swallowed them like Flintstone vitamins—Paxil, Celexa, Prozac, Depakote, and Lamictal. There was physical and mental therapy, shock treatments, and acupuncture. Nothing worked.

Frustrated and incapacitated, and filled with physical pain, he found that his constant impulse to commit suicide ran deeper than before. A move to a bigger house in North Haven, only one mile from his Sag Harbor home, seemed to push the actor further into paralyzing anxiety.

He had talked often of killing himself, and had tried before. It was not unusual for his family to receive messages on their answering machine telling them he'd jumped off the Staten Island Ferry or to find a note scribbled hastily and left on the kitchen table saying goodbye. In 2002 there were a number of documented attempts. Two were on the bridge connecting North Haven to Sag Harbor, and one time Spalding was talked down from the bridge while parading around hyperventilating. The other time, he actually jumped, falling twenty-five feet into the water, where police rescued him. He went sailing with a friend and jumped overboard, announcing that the current should make the decision whether he lived or died.

The day before he vanished, Friday, January 9, 2004, he attended his afternoon psychiatrist appointment, bought his son a book, and was later seen on the Staten Island Ferry, riding back and forth. He left his wallet on a seat and then wandered away. Security escorted him off the boat.

The following day, he was supposed to leave for a skiing trip to Aspen, a Christmas gift from his wife. When he arrived at La Guardia, skis in hand, he was told that his 10:00 AM flight had been canceled because of poor weather conditions. He rescheduled his flight for the following morning, returned to his loft in SoHo, and took his two sons to a matinee screening of *Big Fish*—a story about a dying father's relationship with his son. The trio had a late lunch at an Indian restaurant and then went home. At 6:30 PM, he told his family he was meeting a friend for drinks. That was the last they saw of him. Supposedly, he left his wallet at a friend's house. He called his sons from a pay phone by the ferry terminal before he got on, saying that he loved them.

The last person to see him alive was Hugo Perez, who was riding the same ferry with his girlfriend. The couple spoke to Spalding briefly. He inquired about the protective rope that had recently been installed on the boat. He mentioned to the couple that the view from the ferry was terrific. They didn't think much about it at the time.

The next person to see him was Robin Snead, two months later.

True to form, Spalding did leave a note. His last journal entry, dated December 2003, was written to his wife: "It's an old story you've heard over and over. My life is coming to an end. Everything is in my head now. My timing is off. In the last two years I've had at least ten therapists and all those shock treatments. Suicide is a viable alternative for me instead of going to an institution. I don't want an audience. I don't want anyone to see me slip into the water."

He was a tortured soul who entertained us and, indeed, a great storyteller: vivid and charismatic, relentless at being honest and baring his inner demons, perhaps in the hopes of releasing them. He had a mad scientist's wavy silvery hair, and an instantly recognizable whispery New England accent. He ached to get out of his head the endless ideas and thoughts that never seemed to shut off, to clear his mind and somehow find the peace that eluded him—at least while he was living.

A few months before he died, Spalding was on an upswing. He was seeing the famous neurologist Oliver Saks, was taking a new medication with positive effects, had had an operation that corrected the indentation in his head, and had returned to his pre-accident weight. He'd even started writing again. He rented a room from Performance Space 122 and created a new show, ironically called *Life Interrupted*, which he never completed.

In December, he'd finished crafting the show. The holidays were approaching. He'd bought Kathie a ring for Christmas; she had surprised him with the ski trip. Friends and family were encouraged by seeing him socializing and metamorphosing back into what his son called "Old Daddy."

In the case of many suicidal people, the times—weeks or months—when they appear to be progressing are in fact when they've decided to end their lives. It's a sign many therapists worry about.

"Doctors will tell you that the problem with the recovery of a person in his depressed state is that you have to be very careful," Spalding's wife told *New York* magazine in February 2004. "Because that can mean that they're finally organized enough to carry something out."

Though he wasn't able to complete his final show, friends and colleagues were. In 2006, *Leftover Stories to Tell*, which combined Spalding's unfinished work, notes, and diary excerpts into a play, was performed by Debra Winger, Teri Garr, Steve Buscemi, and Olympia Dukakis, among others, in his honor.

UNEARTHED: Called a suicide magnet, the Golden Gate Bridge is the most popular location in the world to jump from. Since 1937 more than 2,000 people have plunged 225 feet to their deaths, and fewer than 30 people have survived the fall. Ironically, what kills jumpers is not the speed of their fall—about 75 miles per hour—but the impact and the piercing of organs from broken bones as they hit the water; or they drown. If you change your mind mid-jump, entering the water feet first is one of the only ways to survive. Other favorite jumping locations are the Empire State Building, Washington's Duke Ellington Bridge, and Japan's Mount Mihara.

THE OTHER BRIDGE: Built in 1913, the Colorado Street Bridge is a 150-foot-high, 1,467-foot-long concrete structure and scenic gateway into central Pasadena, just northeast of downtown Los Angeles. Upon its opening, it quickly turned into a jumper's delight. From 1919 to 1937, a total of 95 people took a plunge from it.

CAREER HIGHLIGHTS: Though he pushed the boundaries for how he told stories, Spalding was old-fashioned in his creative process. He would write in longhand or dictate his thoughts into a tape recorder from noon until 3:00 PM, then walk around Manhattan or in the Hamptons to clear his mind and gather more thoughts. He wrote only one novel, *Impossible Vacation*, which closely reflected his WASP upbringing and his mother's suicide. Of his numerous one-man shows, most were turned into book form, and a few made their way to film; all were about the various stages of Spalding's life. He appeared in more than thirty-five movies (including *Beaches, Clara's Heart, The Paper*, and *Kate & Leopold*)—and one pornographic flick, *The Farmer's Daughter*, when he was just starting out, in 1973.

Peg, 1930

Peg Entwistle

BORN: February 5, 1908, Port Talbot, Wales
DIED: September 18, 1932, Los Angeles, California
AGE: 24
METHOD: Jumping
DISCOVERED BY: A tourist hiking in the hills
FUNERAL: Held at the W. M. Strothers Mortuary (now demolished), at 6240 Hollywood Boulevard; the funeral was sparsely attended.
FINAL RESTING PLACE: Cremated at the Hollywood Memorial Park; her ashes were sent to the Oak Hill Cemetery, in Glendale, Ohio, where she resides today, next to her father, in section 12, Lot 27, Grave 10.

To play any kind of an emotional scene I must work up to a certain pitch.
If I reach this in my first word, the rest of the words and lines take
care of themselves.

TO A REPORTER AT THE *OAKLAND TRIBUNE*, MAY 1929

For many, memory of a suicide is heightened by the method the suicide uses, the object she chooses to end her life with: a rainbow of pills; a loaded gun; the cool, sharp edge of a razorblade. Some suicides are remembered for their impressive body of work, or for the gene pool they came from. For Peg Entwistle, a young actress from Wales, she's best known for the ironic location of her death.

Dubbed "The Hollywood Sign Girl," thanks to one of the newspapers who covered her death, Peg was the first person to jump off one of the tallest peaks in Los Angeles—the letter *H* of the legendary Hollywoodland sign that reigned prominently on Mount Lee in Griffith Park.

For decades the sign was synonymous with fame and fortune. Like a lover who whispers tenderly in your ear, it seduced young, unknown talent with promises of immortality. Like the sign, from far away Peg appeared radiant and magnetic—a symbol for entertainment and glory. Up close, however, she was dilapidated, worn, and weathered. And though her classic blue eyes, blond hair, and flapper's appeal made her instantly popular among directors, producers, and other industry professionals, at age twenty-four, rejection and abandonment took center stage.

Born Lillian Millicent Entwistle, Peg—who changed her name for theatrical reasons—lived a life that read like a cliché Hollywood tearjerker. Her mother died when she was a child. By the age of eight, she had moved from Wales to London and then to New York with her father and brothers, carrying a suitcase laden with dreams of pursuing an acting career. In 1922, after years of city life, her father, an actor himself and a shop owner, was killed by a hit-and-run driver in Manhattan. Orphaned by the age of fourteen, she was taken in by her uncle Harold. Through family connections, she was cast in a few uncredited walk-on roles in several theatrical plays. A few months later, she found herself in Massachusetts accepting a job from the Boston Repertory Company.

In April 1927, she was wooed by fellow actor Robert Keith. He was a decade older than she, and married twice before. Still, the two became an item, and though he already had a son, which he hid from Peg during their courtship, she married him anyway. A few weeks later, during a dinner party at their home, the police appeared at their door demanding the $1,000 in child support Robert owed his ex-wife. She pulled the money together, but after asking Robert about it, he became violent. He had bad debts. He lied. They fought constantly. The marriage lasted only a few years. Another loss. Another failure.

From 1926 to 1931, Peg worked regularly and was cast in a number of Broadway shows, many of them flops. The most successful was *Tommy*, which had 232 performances, a huge increase from the 12 to 20 shows she was used to doing before a play closed. She toured with national theater companies, changing characters weekly while gaining considerable notoriety and publicity from her solid reviews. The *New York Times* wrote about her twice, offering accolades. While doing *Night of the Barbie* (which closed after only thirty-two performances because the lead, Laurette Taylor, was constantly drunk), Peg became frustrated with theater. The actors were paid for only one week's salary rather than the larger sum they thought they were owed. She decided to make the change to film. And though low on funds, she headed to California, where she was cast in the stage play *The Mad Hopes*. She stayed at the Hollywood Studio Club, a rooming hotel for women, before moving into her uncle's bungalow on 2428 Beachwood Canyon Drive in Hollywood. Though the play was a hit during its tryout period, the show never made it to New York as planned. Still, RKO, then a major studio, signed her to a one-picture deal.

Her first and only movie was called *Thirteen Women*, which she shot in July, and though she had a decent role, when the film was previewed for an audience a month later, many felt it ran too long. As a result, much of her part was cut. Then, only several weeks into her contract, RKO decided not to renew it. She was devastated.

Actors were hit hard by the Great Depression, and Peg was no different. Broke, out of work, and without enough money to return to New York, she was forced to stay in Los Angeles and look for work. None came.

It's unclear what happened during the month of August and the first half of September to cause such a severe breakdown in her. Rumor had it she posed topless for famed pinup photographer Bruno Bernard, the man credited for discovering Marilyn Monroe. Other than that, her career seemed stalled. It was as if she didn't exist.

On Sunday, September 18, feeling rejected and lonely, she told her uncle she was going to the drugstore to buy a book and meet up with friends. Instead, she walked the short distance from her home to the southern slope of Mount Lee, climbed the hill, and stood by the Hollywoodland sign, looking up at the looming letters. She was wearing a dress she'd donned in a Sherlock Holmes play. Knowing this would be her last performance, one without an audience or applause, she folded her coat, placed her shoes and purse at the base of the sign, and scaled the letter *H* using a ladder left behind by workmen who had been replacing burnt-out bulbs.

She stood for a moment, balancing herself, staring out into the tremendous landscape, and then leaped, plummeting the five stories. She hit the dirt hard, then rolled over and over until she fell into a one-hundred-foot ravine.

The following morning a hiker stumbled across Peg's handbag, shoes, and jacket. She followed the trail of scattered, windblown items, which eventually led to the lifeless, rag doll body. The hiker sifted through Peg's purse and found a folded-up note, which simply read: "I am afraid I am a coward. I am sorry for everything. If I had done this a long time ago, it would have saved a lot of pain." It was signed P.E.

Not wanting to identify herself, the hiker made an anonymous call to the police, telling them she'd found the body of a young woman: "I wrapped up the jacket, shoes and purse in a bundle and laid them on the steps of the Hollywood Police Station."

The letter was published in the *Los Angeles Times* the following day in the hope that someone would ID the body. When Peg's uncle saw the initials in the newspaper, he went down to the morgue and put a name to the corpse. The coroner's report lists the cause of death as internal bleeding and multiple fractures of the pelvis, noting that death was not achieved upon impact. Perhaps she lay in a ditch all night, the massive letters towering over her. Perhaps she died just as the sun was creeping up the mountain and illuminating the world. Perhaps it was nothing more than just another day in Tinseltown.

Sadly, no Hollywood story would be complete without a sardonic ending. Two days after her suicide, a letter from the Beverly Hills Playhouse arrived for her offering her the lead role in their upcoming production. She would have played the part of a young woman who commits suicide.

The famous Hollywoodland sign that Peg jumped from in 1932

Many tourists and park rangers claim they've seen a young woman sporting 1930s attire roaming the Beachwood Canyon area. That her signature gardenia perfume scent still lingers, wafting through the night air. In recent years motion-sensitive alarm systems have been installed near the sign to deter people from getting too close, or from "pulling an Entwistle." Still, the alarm is often triggered, the blaring sound emanating through the hills. Though no one is visible, the machine insists someone is standing only five feet away. Not ready to be laid to rest or still trying to play to an audience, she reminds everyone in the vicinity that even a dead starlet can make a comeback.

Though not remembered for her acting, Peg is responsible for single-handedly changing the way the world looks at an already famous landmark. Something that once stood for entertainment and the future now also represents tragedy. Her story is that of a girl who thought she could find solace and salvation in the word *Hollywoodland*. Like the bulbs the workmen had been changing the night she killed herself, Peg burnt out too early. To this day she is the only suicide in the sign's history.

Peg's death certificate

UNEARTHED: For most, jumping is considered an impulsive act. The purposely chosen height offers three things: speed, facility, and certainty of death. Jumpers generally exhibit few warning signs, such as giving away personal belongings, and they have a lower history of prior self-killing attempts. Yet, a huge number of them suffer from schizophrenia and other mental illnesses.

THE HOLLYWOODLAND SIGN: Created as a publicity ploy in 1923 to lure eager homeowners into buying into the Hollywoodland housing development along Beachwood Canyon, the bulb-studded letters cost $21,000 to install and stood at thirty feet wide and forty-five feet high. At night, the four thousand twenty-watt bulbs could be seen for miles and became a visual icon in "LaLa Land." The sign was supposed to reign over the hills for only a year and a half, but eight decades later, it still stands. Maintenance of the grand sign was discontinued in 1939, when upkeep proved too costly. Ten years later, the *land* portion of the sign was removed. In 1978, with the sign in dire need of restoration, celebrities sponsored individual letters for approximately $28,000 each.

DID YOU KNOW: Peg's stepson, three-time Emmy-nominated actor Brian Keith, who was best known for playing Uncle Bill on CBS's 1960s TV hit *Family Affair*, killed himself in 1997, two months after his daughter, Daisy, killed herself—making them three generations of suicides.

CAREER HIGHLIGHTS: Though she was cast in a dozen shows, many closed after only a handful of performances, which filled her with disappointment and added to her unstable emotional state. *The Uninvited Guest* ran for only seven performances. *She Means Business* played for eight. The longest running was George M. Cohan's *The Home Towners*, which had a disappointing sixty-four performances.

Dorothy Dandridge, near tears, at a press conference after testifying in the *Hollywood Confidential* criminal libel trial, September 3, 1957

Dorothy Dandridge

BORN: November 9, 1922, Cleveland, Ohio
DIED: September 8, 1965, Hollywood, California
AGE: 42
METHOD: Overdose
DISCOVERED BY: Her manager
FUNERAL: On September 12, 1965, a private funeral service was held at the Little Chapel of the Flowers, in Las Vegas. Attendees included

Pearl Bailey, Sammy Davis Jr., James Mason, and Sidney Poitier.
FINAL RESTING PLACE: Dorothy requested cremation; her ashes were entombed in the Freedom Mausoleum at Forest Lawn Cemetery in Glendale, California.

Some people kill themselves with drink, others with overdoses, some with a gun; a few of them hurl themselves in front of trains or autos. I hurled myself in front of another White man.
FROM HER AUTOBIOGRAPHY, *EVERYTHING AND NOTHING: THE DOROTHY DANDRIDGE TRAGEDY*

They find her body like this: On the bathroom floor, naked and cold. Her feet are sticking out the door. A blue scarf is wrapped tightly, purposefully around her head. The bathroom tile is like ice. The body is thin and willowy. Her skin is like silk, the color of rich coffee ice cream. She has recently bathed, is perfectly powdered. She's beautiful. But then again, Dorothy Dandridge always was. The handful of antidepressants she forced down her throat in the privacy of her two-story home on Sunset Strip in West Hollywood have done their job. Rigor mortis hasn't set in yet when her manager, Earl Mills, finds her.

Though the suicide shouldn't be a shock to him, the sight of his longtime friend sprawled on the floor is. He had to have known this was possible. Four months earlier, Dorothy handed him a note depicting exactly what she would look like. The letter, only forty-four words long, was filled with instructions. There was no apology. It offered no explanation.

"In case of my death, to whomever discovers it, don't remove anything I have on—scarf, gown or underwear. Cremate me right away. If I have anything, money, furniture, give it to my mother Ruby Dandridge. She will know what to do."

That day, unable to reach his client on the phone, Earl drove to her apartment. When he received no answer from ringing her bell or pounding loudly on the door, he tried the key she'd given him, but the door was chain-locked. Frantic, he retrieved a tire iron from his car and pried the door open. As he ran up the steps, he called out her name. He found her lying on the floor of her bedroom, her hands cupped under her head making it look as if she were sleeping. He covered her with a towel, then a blanket, and phoned for an ambulance. An hour later police filled the apartment. Dorothy's mother arrived next, detectives leading her up the stairs to her daughter's body.

The last few years had been difficult for the well-known actress-singer. She drank heavily, would call friends late at night, babbling on for hours. She was unhappy and bankrupt. She thought her life was going nowhere. Her two marriages were failures. She hadn't spoken to her sister in years. Her eighteen-year-old daughter, who was born with brain damage, had been abandoned on her doorstep by the woman who had taken care of her for the past decade. But she was unable to continue supporting her, and the child would be transferred to a state institution.

Dorothy was the first black woman to grace the cover of *Life* magazine, and the first to receive an Academy Award nomination for Best Actress in a motion picture—and yet, at the time of her death, she had only $2.14 in her bank account.

The actresses who ruled the screen back then were one-name wonders: Marilyn, Liz, Audrey, and Grace. Though she grew up at a time when blacks were forced to use separate bathrooms and sit in the backs of buses, Dorothy seemed determined to integrate a color-obsessed world into a blend of grays. And though whites dominated the cinema screen, she would become as intrinsic to the movie business as Jackie Robinson was to baseball.

She was born in Cleveland, Ohio, in the early 1920s. Her mother, Ruby, walked out on Dorothy's father five months before Dorothy was born, taking her first daughter, Vivian, with her. Ruby cleaned homes during the day and sang and recited poetry at churches and local theater groups in her spare time. After she was born, Ruby met a woman named Geneva, who became her lover and moved in with the three Dandridges. When Geneva wasn't teaching the girls to sing and dance, she beat them.

The four moved to Nashville, where the children sang with the National Baptist Convention and spent the next three years touring churches throughout the southern states—with Geneva at the piano and Ruby handling the bookings and sometimes taking to the stage.

Another move to California landed the Dandridge girls at the Hooper Street School, where they met classmate Etta Jones. The trio soon became a well-known act, touring and performing at places like the famous Cotton Club, where she met Cab Calloway, Bill "Bojangles" Robinson, and the well-known tap dancer Harold Nicholas.

From the late 1930s to early '40s, Dorothy performed in short films while headlining at major clubs and hotels. Though her voice oozed glamour, racism ran rampant then, and she was often turned down for roles because she was black. In 1942 she married Harold Nicholas, a selfish womanizing playboy, and the two set up house in a Hollywood Hills mansion. She was earning an unheard-of $100,000 per movie, wore a white beaver fur coat, drove a white

Thunderbird, and dripped in jewels. To the outside world, she had it all.

When she went into labor a year later, Nicholas insisted it was a false alarm and went to play golf instead of taking her to the hospital. She waited hours for him to come home. Eventually she gave birth to their daughter, Harolyn, and hoped things would get better. She took some time off to be a mother, intent on giving her daughter everything she never got: two loving parents and a family without abuse.

As time passed, it became clear that her daughter wasn't well. She didn't walk or talk, and she cried all the time. Doctor after doctor each said the same thing: lack of oxygen to the child in utero had caused mental retardation. Dorothy blamed herself. And following in her mother's footsteps, she took her child, left her husband, and filed for divorce. She returned to performing, the only thing she knew how to do. Though she suffered from horrific stage fright, and hated the nightclub scene—the smoking, the booze, the groping from drunken men—she revamped her act. Soon she was being booked in "whites only" nightspots and hotels. Though she was the first black woman to perform at Manhattan's Empire Hotel and the Waldorf Astoria, she was not allowed to stay in one of their rooms. She was also not permitted to speak with patrons, walk through the lobby, or ride in the elevator. One hotel drained its pool so she couldn't enjoy that amenity.

Dorothy fought hard for each and every job she got: a jungle queen in *Tarzan's Peril*, an athlete's girlfriend in *The Harlem Globetrotters*, a young schoolteacher in Alabama for a big-budget MGM film called *Bright Road*. All the while, she continued to perform in top clubs and made numerous guest television appearances.

Her big break came in 1954, when she was cast as the lead role in Oscar Hammerstein's *Carmen Jones*, an adaptation of Bizet's legendary opera *Carmen*, and the first all-black musical film. She played opposite Pearl Bailey and Harry Belafonte—and started dating the film's director, Otto Preminger. She was nominated for an Academy Award for Best Actress, and though she didn't win—Grace Kelly did that year, for *The Country Girl*—she was the first African American to be acknowledged in that category.

After *Carmen Jones*, a string of other projects came her way, none of which brought her the same kind of recognition; nor were they roles she wished to play. Frustrated at being cast in only "black" parts—slaves, maids, and mammies—she passed up the part of Tuptim in *The King and I*, one of her biggest mistakes; the movie was a huge success. Eventually she signed on to do the film version of George Gershwin's musical *Porgy and Bess*, which also starred Sidney Poitier and Sammy Davis Jr. There she was reunited with ex-boyfriend Otto Preminger, the film's director.

June 1959 was a big month for Dorothy. She was filming *Porgy and Bess* and married Jack Denison, a white restaurant owner who had pursued her relentlessly and who many insisted was interested only in her money. They proved to be right. He soon took over her career, shut out her family, and commandeered her life. He was verbally and physically abusive. He took away her pride, her friends, and her money. She got pregnant and, afraid of having another disabled child, had an abortion. The couple divorced three years later.

The Murder Men was her final film role and marked the point at which her life seemed to fall apart completely. Her drinking escalated. She was involved in eight lawsuits from creditors. She was out of control, out of money, and out of work. To lift her spirits, a doctor prescribed an antidepressant called Imipramine. She started doing the nightclub circuit again, trying to return to her roots, but many critics felt her sparkle had gone.

A few years later she had, too.

The morning of the day she died, Dorothy spoke to her friend Geri Branton. She also cancelled an appointment she had with her doctor, who was going to put a cast on her right foot, which had a tiny fracture. She had come home the night before from a paid singing gig at a club, and the following day she was due to leave for a New York engagement. Her bags were packed. She had more films and performance jobs lined up over the next several months than she'd had in years.

It seemed as though nothing came easily to Dorothy. Her entire life was full of sadness and disappointments. Even her simple request to be cremated right away ended up taking months.

During the autopsy, some specialists thought an embolism caused by bone marrow escaping into her bloodstream from the tiny break in her right foot was the cause of her death. But an eighteen-page coroner's report later dismissed this idea, ruling "Hollywood's first authentic Black sex symbol had died from acute drug intoxication due to an overdose of Tofranil."

In 1999, Halle Berry portrayed the five-foot-five, honey-colored actress for HBO's *Introducing Dorothy Dandridge*. Berry told the *New York Times* that in her mind, Dorothy's death was a suicide: "I am inclined to believe, especially after playing her and looking at some things that have happened in my life, that she was so discouraged that she killed herself. Even though she had packed her bags, I think, somehow, she had reached the end. She was just tired."

The Los Angeles Police taking Dorothy's body out of her residence

UNEARTHED: Created and given out in the 1950s, Imipramine was the first tricyclic antidepressant, and it seemed to cause as many problems as it solved. Though originally created as a mood elevator with some positive results, it also induced and exacerbated psychosis and caused a high rate of mania, especially in people with preexisting bipolar disorder. To this day, Imipramine is still prescribed, and is a favorite among some doctors.

DID YOU KNOW: The most common pills used for suicide are tranquilizers and sleeping pills. At 37.8 percent, overdosing (or poison) is the most common suicide method for women. Little cleanup and easy attainability make pills and poisons likable choices—accounting for 38 percent of female, and 15 percent of male suicides. A pill overdose is relatively painless, and also offers the highest chance for survival if the suicide is discovered and taken to the hospital in time to have her stomach pumped.

CAREER HIGHLIGHTS: From 1935 to 1961, Dorothy appeared in more than thirty-five films, and sang on *The Ed Sullivan Show, Toast of the Town*, and *The Colgate Hour*, among others. She also penned a memoir, *Everything and Nothing: The Dorothy Dandridge Tragedy*. Completed right before her death, the book was released five years later, in 1970, and reissued again in 2000.

David Strickland with his best friend and sitcom-mate Brooke Shields at the 25th Annual People's Choice Awards on January 10, 1999, two and a half months before he hanged himself

David Strickland

BORN: October 14, 1969, Glen Cove, Long Island
DIED: March 22, 1999, Las Vegas, Nevada
AGE: 29
METHOD: Hanging
DISCOVERED BY: Housekeeping
FUNERAL: His funeral was held at the Forest Lawn Memorial Park, in Glendale. Cast members from *Suddenly Susan* were there: Brooke Shields, Kathy Griffin, and Judd Nelson, along with David's friends Juliana Margulies and Tiffani-Amber Thiessen.
FINAL RESTING PLACE: David was cremated, and his ashes were given to his parents.

I'd like to be 30, 38, 39 and doing something really great and sort of have this look back and say, "He's worked consistently."
A FEW WEEKS BEFORE HIS DEATH, TO AN ASSOCIATED PRESS REPORTER

The Los Angeles courtroom is quiet as the handful of lawyers and some friends wait for actor David Strickland to appear for his hearing. Back in October, the *Suddenly Susan* actor was arrested for cocaine possession. He pleaded no contest, was put on a three-year probation, and ordered into a rehab program. Today's court appearance is supposed to be a progress report.

The group continues to wait, frustration and anger filling the courtroom like a thick cloud of smoke. He's not going to show. This becomes painfully clear as the minutes tick by.

Two hundred and seventy-five miles away, at the Oasis, a sleazy Las Vegas motel, a phone rings in room number 20. When there's no answer, a motel employee knocks on the door and then, as is standard policy, uses a key to enter the room. She doesn't have to open the door far; a foot or two is all she needs to see the room's occupant hanging from the ceiling beam, the bed's king-size sheet tied harshly, tightly, around his neck. His body, dressed in jeans, a khaki shirt, and black high-top sneakers, is motionless. The room is silent, still. A chair is next to his body. Empty cans of beer, purchased at a local convenience store, are lined up like tin soldiers in perfect formation on the edge of the nightstand. That's when the screaming starts.

"I was working at the front desk when one of our employees came running into the room hollering and crying," says Peter Napoli, the motel manager, the second person to see David hanging. "It's the first thing you see when you open the door. It was a horrible sight. It took several seconds to register what was happening." Though Napoli doesn't think about the incident now, even when passing by the room, he still prays for the actor every day. "Some can't handle success, that's my assumption. They get into drugs, that's the business, and all the problems that go along with it," he adds. "There's no good reason for them to take their lives, but people do it every day. The casinos just know how to keep it quieter. That's not the kind of place we are. We're a small place for people to hide out. We get the good, the bad, and the ugly."

No note or drugs were found in the room or on David. Police searched for a car in the parking lot and came up empty, which made it unclear how he got to the motel, or where he was before he arrived at the strip club two miles from the motel. The police phoned the coroner's office, which in turn sent someone to perform the medical investigation. "There are five jobs of that investigator," explains Mike Murphy, of the Clark County Coroner's Office: "Determine cause of death, determine method, identify the body correctly, notify the family, and protect the property of deceased person." His investigation started at 12:30 PM and lasted several hours.

The night before had started out like a poorly planned fraternity party—a fast drive into Vegas and a strip club with fellow AAer and rehabber, comedian and friend Andy Dick. The duo was seen at the Glitter Gulch strip bar around 1:00 AM, where dancer Kimberly Braddock gave David a private lap dance. Some say he took her back to the Oasis. Others claim a woman was waiting for him in one of the motel rooms. Regardless, he left one room and proceeded back to the motel office to request a second. For sixty-four dollars, he was given another one, which he stayed in from around 3:30 to 10:00 AM.

When he missed his court hearing, Brooke Shields, his costar and dear friend, hired a private investigator to help in the search. She was one of the few who knew how tormented he was. A bipolar alcoholic, he'd had trouble with drugs and had gone to rehab. Though people believed he was doing better, he was depressed and had stopped taking his lithium. As with Spalding Gray, friends and industry peers also felt his career was on an upswing. *Suddenly Susan* was doing well—returning for its fourth season—and though most of his part in *Forces of Nature* had been left on the cutting room floor, the film was number one at the box office. Also, he was dating *Saved by the Bell* and *Beverly Hills 90210* actress Tiffani-Amber Thiessen, and talked of wanting to direct an episode of *Suddenly Susan*. He seemed sober, busy, and happy. No one knew he wanted to kill himself. But bipolars have a high rate of suicide, and the Clark County coroner's report says that marks from previous suicide attempts were found on Strickland's body.

"Almost everyone who lives here is not from here," says Murphy. Clark County has a population of 2.2 million, with 150,000 tourists per day, and the Coroner's Office sees 300 to 320 suicides a year. "There's a very little home networking system. People come here for their last hoorah with the intention of killing themselves."

Born in Glen Cove, Long Island, David moved to New Jersey with his family before settling in the Pacific Palisades, a suburb of Los Angeles, where he finished high school at Pacific Palisades High. Rather than attend college, he decided to pursue acting, something he had never considered before. He joined a theater company, where he focused on comedy sketches and worked on fifty student films, gaining the acting experience he had been lacking.

After a handful of bit parts, a few recurring roles on sitcoms such as *Sister, Sister* and *Mad About You*, his big break came when he was cast alongside Brooke Shields, Judd Nelson, and Kathy Griffin in NBC's *Suddenly Susan*. The show revolved around a fictional San Francisco–based magazine, *The Gate*. David played the likeable, quick-witted Todd Stites, the magazine's rock critic. His all-American collegiate look—attractive, clean-cut, with piercing blue eyes—made him commercial and castable. His easygoing manner also helped him blend seamlessly into the celebrity world.

Just as David didn't materialize in the courtroom the day of his hearing, his character on *Suddenly Susan* didn't appear in the third season's final episode, either. In the episode, "Todd Stites" simply doesn't show up to work one day. When Brooke Shields's character, Susan, phones him, his pager vibrates on his office desk. Confused and worried, she spends the rest of the episode searching for him, in the process, finding out that he did a number of good deeds throughout his life.

Toward the end of "A Day in the Life," which aired on May 24, 1999, and was dedicated to him, police enter the magazine office. Eager staffers inquire about their friend. No real answer is given. Tribute is paid by the characters, who share personal experiences about Todd throughout the episode. His favorite song, Fat Boy Slim's "Praise You," is played outside in the street as Susan and her co-workers sit in a circle praying for his well-being.

The last line of the show is delivered by Shields: "It's the things you had a chance to say every day and didn't that you end up regretting." *Suddenly Susan* lasted only one more season, going off the air in 2000. And David never got to experience being thirty-nine, thirty-eight, or even thirty, as he had hoped.

UNEARTHED: Hanging is the third most common form of suicide, accounting for 16 percent of males and 13 percent of females, and most will choose a chair or a ladder to complete the task. "If you can tie it, it's been used," Murphy says of the materials used for hanging. "From clothing to electrical cord from the lamps in the room, to cords from a computer, ropes, cables, bed sheets, shoestrings, belts—people can be extremely creative." Since the neck is rarely broken—one needs a drop of six feet or more for that to occur—most people who hang themselves die from choking. The face color can range from pale to cyanotic blue, depending on whether much blood was trapped in the head. If the ligature puts only enough pressure on the neck to close the jugular veins but not the carotid artery, a swollen, blue, blood-congested face is the result.

DID YOU KNOW: The most commonly used method to kill yourself in a hotel room is with a gun. "Firearms is the most prevalent method," says Dr. Paul Zarkowski, a psychiatrist in the Department of Psychiatry and Behavioral Sciences at the University of Washington. "It's a very effective, easy way to end your life. Mostly because there's no object you need to find to hang yourself with."

ANOTHER VEGAS SUICIDE: There are more hotel rooms in Las Vegas than any other city in the United States. And in 2001, twenty-five-year-old Justin Pierce, best known for his portrayal of a skateboarding punk in the 1995 film *Kids*, hanged himself in the Vegas Bellagio resort, only three miles from David's motel room. Though the hotel was much fancier, the method was the same. Pierce left two suicide notes A new father and husband, he had recently finished filming *Next Friday* with Ice Cube when he checked himself in and was later found by hotel security.

CAREER HIGHLIGHTS: Of the ninety-three episodes shot for *Suddenly Susan*, David taped seventy-one of them. He got his start playing a driver in 1994's *Postcards from America*, which starred Michael Imperioli. The following year, he played a homeless man in *Object of Obsession* and worked on *Phobophilia: The Love of Fear*, starring the magicians Penn and Teller. He also appeared in a handful of sitcoms, in a few recurring roles. A lead part came when he accepted the role of a pizza delivery guy in the independent film *Delivered*, before moving on to the commercial success of *Forces of Nature*, starring Sandra Bullock and Ben Affleck, which was his last film project.

Actors Unearthed

THE FIRST TV SUICIDE: In 1974, one month shy of her thirtieth birthday, newscaster Christine Chubbuck uncharacteristically opened her show on WXLT, *Suncoast Digest*, with a news story. Seated at the anchor's desk, a .38 revolver stored in a bag of puppets beneath her, she told viewers she'd soon be dead. She then covered three national news stories and a local restaurant shooting from the day before. When the footage of the restaurant shooting jammed, she improvised: "In keeping with Channel 40's policy of bringing you the latest in blood and guts, and in living color, you are going to see another first: an attempted suicide." She quietly, systematically, retrieved her gun, raised it behind her right ear, and with a shaking hand, pulled the trigger. She was still on the air as she fell forward, blood leaking out from her skull, the potent smell of smoke filling the air. Her body slid to the floor as the technical director faded to black. A public service tape played, and then a movie for viewers to watch as co-workers tried to process what had transpired. Chubbuck died fourteen hours later, at Sarasota Memorial Hospital.

Heavy with depression, she often talked of killing herself and had attempted to do so via an overdose five years earlier. A week before her suicide she told a night news editor that she had purchased a gun and joked about killing herself on air. Three days before her on-air death, she was told by her director to cover more gruesome stories. During the weeks that followed her death, WXLT aired reruns of the TV series *Gentle Ben* in place of Chubbuck's show. Chubbuck was cremated, and her ashes scattered into the Gulf of Mexico.

CREMATION · Eerily, each actor highlighted in this book was cremated and allowed a proper burial. This wasn't always the case. In early times, suicides were often denied funeral rites and burial in a church cemetery. When Archduke Rudolf, the heir to the throne of the Austro-Hungarian empire, killed himself in 1889, the medical bulletin declared evidence of "mental aberrations" so that Pope Leo XIII would grant him a religious funeral and burial in the imperial crypt.

Before any arrangements can happen, a mandated twenty-four- to forty-eight-hour "waiting period" is usually required. The deceased is kept in a temperature-controlled refrigeration unit until the cremation can be performed.

It works like this:

- The casket or container is placed in the cremation chamber, where the temperature is raised to between 1400 and 1800 degrees Fahrenheit.
- When performed individually, cremation of the average person can be completed within two to two and a half hours.
- All organic matter is consumed by heat or evaporation.
- Any leftover residue, such as bone fragments, are removed from the cremation chamber and processed in a machine until they resemble coarse sand, light gray in color.
- The next resting place is usually an urn.
- If you want to watch, you may. Arrangements can usually be made, through a cremation authorization form, for relatives or representatives of the deceased to witness the cremation.

Since cremation is an irreversible process and eliminates the ability to determine exact cause of death, many states require that each cremation be authorized by the coroner or medical examiner.

Five
Musicians

Some might argue that there is no one more beloved, admired, or mimicked than the rock star. Who hasn't stood on their bed playing air guitar while singing along to a favorite song? As with actors, we are in awe of their godlike image. Their bad behavior is expected, applauded, and overlooked because of the strength of their work, their brilliant lyrics, their magnetic qualities. We don't just adore them, we often want to be them.

Sadly, their "live fast, die young" mentality has often been taken too far—to an early grave. They perform hard, live harder, and party dangerously close to the great self-destructive abyss. In the early seventies, a rash of high-profile, greatly adored, and extremely accomplished musicians all OD'd within a year of each other, though their deaths were accidental: Jim Morrison, Janis Joplin, and Jimi Hendrix—each dead at age twenty-seven. Each left a legacy, along with an insatiable appetite from their fans.

American popular music has gone through a great number of transitions since the middle of the twentieth century. In the goodie-goodie, utopian 1950s, Fats Domino found his thrill on "Blueberry Hill," Debbie Reynolds and Carleton Carpenter hit the charts with "Aba Daba Honeymoon," and Doris Day shrugged it off with "Que Sera, Sera (Whatever Will Be Will Be)." Music then changed from the saccharine tunes of that decade to the free love, antiwar, flower power songs that became synonymous with the mid- to late 1960s. There was Motown, too, which brought with it its own inspiration. Motown songs acknowledged pain and hardship, but were optimistic about someone being there to lend an ear and a hand. The Five Stairsteps soothingly promised that things were "gonna get easier" with "O-o-h Child," and Diana Ross and the Supremes swore that if we needed someone, they'd be there in a hurry with "Ain't No Mountain High Enough." A huge shift happened in the 1970s. British punk rock pulled up a loud, radical, and volatile seat next to the happy roller-skating beat of disco. The two genres clashed instantly, and rarely did you find a listener who appreciated both.

As the times changed, so did the music—and with it, the lyrics and expressions artists wanted to explore, create, and share. Punk brought with it gritty, angry, harsh performers. Sid Vicious and Darby Crash, poster boys for that era, killed themselves by the ages of twenty-one and twenty-two, respectively. As punk turned into post-punk, and post-punk morphed into grunge, a slew of new artists, also self-destructive and clearly addicted to one form of drugs (or a host of them), emerged. Their image was different. They were physically meeker, sadly thin, and visibly unhappy and dysfunctional: Nick Drake, Ian Curtis, Kurt Cobain, and Elliott Smith. All young. And between 1974 and 2003, they were all dead before the age of thirty-five from suicide.

Gender, too, seems to play a huge role in this suicide group dominated by men. Wendy O. Williams and Susannah McCorkle are two of the most prominent, but are hardly remembered, as women barely exist in the sea of musical testosterone. Neither is mourned each year by throngs of fans, as are Kurt and Elliott. There are no biopics in the works for them as there are for Nick Drake and Darby Crash, no books or documentaries written about them. In the saddest terms, they are overlooked, forgotten about when compared to the Sex Pistols' Sid Vicious or The Temptations' Paul Williams—gods to many in the music industry, kings and saviors to their fans.

Albert Ayler, a 1960s avant-garde jazz saxophonist and singer, disappeared in 1970 and was found days later in Manhattan's East River. Rumors still circulate that this alcoholic and coke user was murdered, mostly because of his involvement with the Black Power Movement. But as others stepped forward to say he'd been depressed, police called the act a suicide.

In 1986, Canadian composer and singer Richard Manuel, a chronic substance abuser, hanged himself in a motel room during The Band's reunion tour in 1986. In 1993, guitarist Douglas Hopkins, having been fired from the Gin Blossoms due to his drinking and unprofessional behavior, was pushed over the edge with the band's subsequent success without him. The depressed alcoholic sneaked out of Phoenix's St. Luke's Hospital during an intake consultation, purchased a .38, and shot himself. And in 1999, The Sound's lead singer-songwriter, Adrian Borland, who was close to finishing a solo album, threw himself in front of a train right before the record was to be completed.

More recently, in 2007, a note was found taped to the front door of the home of Boston's lead vocalist Brad Delp: "To whoever finds this I have hopefully committed suicide. Plan B was to asphyxiate myself in my car." Delp's fiancée had called the police after discovering Delp's car with the dryer hose attached to it. A second note was found on the door at the top of the stairs directing police to the master bedroom. A third warned of carbon monoxide. Tape sealed the bathroom door from the inside, where Brad was found encapsulated in blue-gray smoke from two lit charcoal grills. His head rested on a pillow; a note was attached to his shirt. It read: "Mr. Brad Delp. J'ai une âme solitaire. I am a lonely soul."

Based on bleak glamour, these rock stars and their bands sprouted cultish followings. Each singer's death was thought by many as central to the development of punk, grunge, and the pop culture music community.

What stands out most with this group is how shockingly murky their deaths are when compared with those of the poets featured in this book. For the latter, suicide was the clear aim: Sylvia Plath sealed off her kitchen and put her head in the oven; Ernest Hemingway and Hunter Thompson shot themselves. With the exception of Ian Curtis, this is not the case with the musicians, who perhaps in an attempt to self-medicate and dull the pain ended up killing themselves with their addictions, which, for many of them, came along for the ride throughout their lives—an unwanted guest who sat at their tables, took up space in their homes and lives. Which begs the question: How much is suicide addiction-related?

With each of their deaths there was the hiring of private detectives, additional police investigations, elaborate autopsies by medical examiners, and countless books probing cover-ups and conspiracies. There were also biopics, documentaries, biographies, and autobiographies by everyone from bandmates to family members and close friends, all attempting to share a personal look or explain the deaths.

Ian Curtis performing onstage at Lyceum, UK, February 29, 1980

Ian Curtis

BORN: July 15, 1956, Old Trafford, Manchester, England
DIED: May 18, 1980, Macclesfield, Cheshire, England
AGE: 23
METHOD: Hanging
DISCOVERED BY: His estranged wife
FUNERAL: Family and band members came together to mourn Ian's death at the Macclesfield Crematorium, where his body was cremated and a memorial stone was laid. The group then went to his wife's parents' house to pay their respects.
FINAL RESTING PLACE: His ashes were buried in Macclesfield Cemetery. The inscription on his memorial stone—"Love Will Tear Us Apart"—was chosen by his wife, Deborah, as a tribute to Joy Division's hit song.

Instead of just singing about something you could show it...
IN AN INTERVIEW WITH *NORTHERN LIGHTS* MAGAZINE, NOVEMBER 1979

Birmingham University, 1980. In a baritone voice as deep and haunting as Jim Morrison's, sang Ian Curtis, front man of Joy Division, whose short-lived career made a major impact on the British postpunk movement.

Wearing a boyish look reminiscent of a British lad in the 1970s, in a freshly pressed gray shirt and clean black slacks, he grasped the mic with both hands, as if to prevent himself from collapsing. With his eyes shut, it appeared he was close to passing out. Sometimes he did. When they were open, his eyes were intense, wide, as if surprised. His movements were oddly militaristic: quick and sharp, a robot running in place. Other times they were spastic, his arms flying out like a bird flapping its wings, ready to take flight. When the lights flashed too quickly or the drums beat too heavily, or when the tempo was too fast or he'd had too much to drink before going onstage, or if he was stressed, sleep deprived—and what rocker isn't?—then his epilepsy would emerge, and a seizure was inevitable.

It was not unusual for Ian to seize onstage and have to be carried to the dressing room by his three band members. There he'd break down in sobs. Fans couldn't tell if the twitching, the flopping on the floor, the gulping for air, were part of his act or part of his condition.

Two weeks after the Birmingham incident, he twitched and shook for the last time when the noose around his neck pulled at his carotid artery, cutting off his blood flow and killing him within minutes.

Around 5:00 AM, the sound of Iggy Pop's voice carried throughout Ian's small house in Cheshire, on 77 Barton Street, like a comforting lullaby. On the table sat an empty coffee mug and a whisky bottle, a photo of Ian's thirteen-month-old daughter, Natalie, and his wedding picture. His wife, Deborah, the last person to see him alive, hours earlier, was the first to find the body.

A man of words, Ian had left a long, intensely intimate letter for her, next to his turntable, on which played Iggy Pop's *The Idiot*: "At this very moment, I wish I were dead. I just can't cope anymore." His handwriting was all in capital letters, the same way he wrote his songs. He expressed love for Deborah and their daughter and the hatred he felt for Annik Honoré, the young Belgian woman he'd met while on tour in Europe, and with whom he was having an affair.

The evening of his death, Ian returned to his home in Macclesfield hoping to reconnect with Deborah and beg her to drop the divorce suit she'd filed. He drank. They fought. She didn't want to change her mind. He drank more, then asked her to leave him alone, saying he planned to catch the train to Manchester in the morning. The following day the band would be leaving for America, on their first major tour. It was everything he'd always wanted.

He spent the evening watching Werner Herzog's 1977 film *Stroszek*, about a Berlin man who is released from prison, goes to America searching for a new life, and ends up killing himself instead. He put on Iggy's album, wrote the note to Deborah, walked into the kitchen, climbed on a chair, looped a long piece of rope around his neck, and hanged himself.

Born to working-class parents, Ian grew up in Hurdsfield, a small industrial village in Cheshire, England. He loved poetry and reading. His obsession with pop culture and rock stars such as David Bowie and Lou Reed started early. They were his closest self-created friends—the Velvet Underground, The Stooges, Roxy Music—his family.

As a teen, he and his high-school pal could be found doing drugs they'd stolen from the homes of the elderly, whom they were supposed to be visiting as part of a social services program. This proved nearly fatal when, in 1975, he accidentally overdosed on Largactil, a drug used to treat schizophrenia. After having his stomach pumped, he dropped out of school and took a job as a civil servant. A few months later, he married Deborah. Both hated their jobs and lived more on Ian's dreams of stardom than they did from their paychecks.

His life changed significantly when he went to a Sex Pistols concert. Intimate shows meant fans became familiar with one another. At the handful of shows Manchester's prime punk club was hosting, it was inevitable that Ian would cross paths with fellow Sex Pistols fans Bernard Summer and Peter Hook, who were looking for a lead singer for their band. Stephen Morris became their fourth member, and soon the band Warsaw—a name pinched from David Bowie's song "Warszawa"—was formed. Upon learning that another group with a similar name existed, they crafted a new name: Joy Division.

Most nights Ian could be found in his office—dubbed "The Blue Room"—with the door shut, hunched over a desk swimming in books and living on coffee and Marlboros. An avid reader of Dostoevsky, Kafka, and J. G. Ballard, he found inspiration in their dark, chilling works. These characteristics crept into his lyrics, which were filled with dread, despair, and heartbreak.

The years 1978 and 1979 carried a whirlwind of life-altering events for him. Joy Division got a manager, a record deal, bookings, and a quick cult following. During this time, he was also diagnosed with epilepsy, his daughter was born, and *Unknown Pleasures*, the band's first album, was released. Later they debuted "Love Will Tear Us Apart," soon a huge hit, and became post-punk darlings.

At rehearsals he would direct the band, listen to the music they played, and teach them the songs he'd written the night before. Sometimes he would be inspired enough to create lyrics on the spot.

As the band grew more famous, Ian's moods, rages, drug and alcohol use, and epilepsy worsened. He'd have onstage seizures that left him humiliated, frustrated, and angry. The barbiturates he was taking made him groggy; his other medication—which he would take too much of—caused horrible mood swings. He was also heavily involved with Annik Honoré, and the guilt of cheating on his wife was causing a constant drowning-like sensation.

On April 7, 1980, he attempted suicide by overdosing on his epilepsy medication. Surprisingly, no one took his attempt seriously. His band members were small-town boys, eager and ambitious, dreamers who'd spent two years playing dumps and dive bars. They were ready for the success they'd been promised. So the shows went on, the group kept working; they shot a video for "Love Will Tear Us Apart" and started preparing for their first American tour.

As his love affair with Annik continued, his depression mounted. The epilepsy was ruining his performances. The medication was ruining his moods. And so he found himself, in mid-May 1980, standing on a chair in his kitchen, with Iggy's voice in the background, birds singing, the sun rising as he took a final step forward and entered permanent darkness.

In the aftermath, friends and record executives echoed one another's sentiments of regret: for the warning signs they missed, and the missed opportunities to intervene.

"I think all of us made the mistake of not thinking his suicide was going to happen. . . . We all completely underestimated the danger," said Tony Wilson, who ran Factory Records, Joy Division's record company, during an interview for *Spin* magazine. "We didn't take it seriously. That's how stupid we were."

His bandmates were also guilty. Never once did they sit down to truly listen to his lyrics until he was dead. Once they did, they realized how much pain he was in.

"People constantly ask, 'Why did he kill himself?' " Ian's daughter wrote in an article for Guardian.co.uk in 2007. "To me it seems obvious because he was really depressed. Bernard [Sumner, Joy Division

Ian's gravesite at the Macclesfield Cemetery, UK

guitarist] told me that my father used to drink before performing, which may explain his on-stage fits, because alcohol is a seizure trigger. Seizures can also be triggered by flashing lights, lack of sleep and stress. Ian's lifestyle and the tension caused by the disintegration of his marriage would not have helped. He did the best he could; he was just very ill."

Ian constantly blurred the lines between art and reality, performance and illness, intended dance movements and epileptic fits. His deep voice and even deeper, darker, and sensitive lyrics endeared him to fans. But it was his suicide that propelled him into "depressed singer, doomed rock star" immortality.

UNEARTHED: "Hangers" are the group most apt to leave suicide notes. The most common knot used is the slipknot, which is usually placed at the side of the neck. Only four pounds of pressure is required to close off the jugular vein, and eleven pounds of force closes off the carotid artery. Specialists will look for a *V*-shaped bruise when a rope, electrical cord, or belt is used, to tell the difference between a suicide or a murder (which is indicated by a bruise in a straight line). Approximately two thousand hangings happen in England yearly. (That number almost doubles in the United States.) A UK study performed in 2002 showed that of the 162 suicides by hanging, 106 were done at the person's home. A rope or cord was the most commonly used device, accounting for 79 hangings. Household clothesline remains a favorite, except among the wealthier population, who opt for electrical cords.

DID YOU KNOW: In the past five years, Korea has had the highest rate of suicide in the world. Because owning a gun in South Korea is illegal, hanging is the most common form of suicide there.

CAREER HIGHLIGHTS: Two deaths occurred the night of Ian's suicide, his and Joy Division's. An early pact made by the band stated that if any member decided to leave and the others wanted to stay, they would change the name of the band. Keeping their word, New Order was formed by the remaining members. Like the Doors and The Fall, Joy Division took their name from a memoir, in their case, Yehiel De-Nur's *The House of Dolls*, which describes how women in Nazi concentration camps were used as sex slaves.

Two months after his death, the band's last album, *Closer*, was released; it is considered one of the most important albums in the post-punk movement. Their first album, *Unearned Pleasures*, became a mouthpiece for how people were feeling, a statement about their environment and their daily lives. The first and only time Joy Division performed "Ceremony" was during their Birmingham University concert. Later, the renamed band New Order recorded and released the song as their first single. Many of their initial songs were written as tributes to Ian, such as "I.C.B" (Ian Curtis Buried). The last song Ian performed onstage was "Digital," which can be found on *Still*, a compilation album.

Though hardly mainstream in its day, the band still has an enormous cult following, thanks to the number of recent books, biopics, and documentaries, including Deborah Curtis's memoir *Touching from a Distance*, the book on which the 2007 film *Control*, about Ian's life, was based.

Th BoddAH *pronounced*

Speaking from the tongue of an experienced simpleton who obviously would rather be an emasculated, infantile complainee. This note should be pretty easy to understand. All the warnings from the punk rock 101 courses over the years, since my first introduction to the, shall we say, ethics involved with independence and the embracement of your community has proven to be very true. I haven't felt the excitement of listening to as well as creating music along with ... reading and writing for too many years now. I feel guilty beyond words about these things. For example when were back stage and the lights go out and the manic roar of the crowd begins. It doesn't affect me the way in which it did for Freddy Mercury who seemed to love, relish in the love and adoration from the crowd. which is something I totally admire and envy. The fact is, I can't fool you. Any one of you. It simply isn't fair to you or me. The worst crime I can think of would be to rip people off by faking it and pretending as if im having 100% ...

Kurt's controversial suicide note, 1994

Kurt Cobain

BORN: February 20, 1967, Aberdeen, Washington

DIED: April 5, 1994, Lake Washington, Seattle, Washington

AGE: 27

METHOD: Gunshot and possible overdose

DISCOVERED BY: An electrician

FUNERAL: Aside from a vigil held at Viretta Park, a private memorial was held at the Unity Church of Truth, where Nirvana bandmate Krist Novoselic delivered the eulogy.

FINAL RESTING PLACE: Kurt's ashes are sprinkled in a number of places. Courtney Love spread a handful around the willow tree in their front yard at 171 Lake Washington Boulevard, in Seattle. Buddhist monks blessed the rest of his ashes, then used a fistful to make a Tsa Tsa memorial sculpture. During a memorial five years after his death, a clump of his remains was deposited into the Wishkah River, in Washington State, and in McLane Creek, in Olympia, by his daughter. The rest resided with Courtney, who, in 2008, claimed someone stole them from her closet.

On April 10, over a loudspeaker, Courtney Love reads from her husband's suicide note as part of the eulogy. You can hear the anguish in her raspy, jagged, harsh voice as it cracks and breaks before the six thousand-plus fans who worshiped the King of Grunge.

The public vigil for Kurt Cobain is happening at Viretta Park in the Seattle Center, only minutes from his and Courtney's home. Graffitied personal notes from fans cover the benches and the stairway leading to the top of the park. Flannel shirts are set on fire as an homage. Other fans run into the fountain, as if to baptize themselves as Nirvana's songs blare from the speakers. The crowd morphs into a sea of flowers and lit candles, Tears and moans can be heard. Some mourners yell, others comfort. All are devastated by the loss. Prerecorded messages by Krist Novoselic and Courtney are played before she materializes to read portions of his suicide note, adding in her own comments as she struggles through the service. When his note describes how it would be a crime for him to stay in a business he no longer feels passion for, she says, "No, Kurt, the worst crime I can think of is for you to continue to be a rock star when you fucking hate it and just fucking stop." When he expresses the need to stop being a music symbol, she follows with, "Then don't be a rock star." So angry with him, she asks the crowd to yell out obscenities. "Say 'asshole,'" she begs. And they do. You can hear their sorrow and feel their pain, their need to grieve with and for her in their chants.

When she gets to the part where he insists it's better to burn out than fade away, she tells the crowd not to listen. "Don't remember this," she warns, "because it's a fucking lie. We should have let him have his numbness instead of trying to strip away his skin."

You can almost picture the tears rolling down her cheeks, the snot dripping from her nose, hear how broken she is inside, from the sadness and devastation she wears onstage and from the words pouring out her lips. Then her voice fades away. Later, she appears dressed in black, and as a special thanks to those who stayed till the end of the service, she gives the audience items of Kurt's clothing.

Hours earlier she and seventy of his close friends and family gathered at a private memorial for her husband to say goodbye.

In the early 1990s, Kurt Cobain became synonymous with teenage angst. His stringy, dirty hair, ripped jeans, frail, skinny body, and scratchy voice filled with pain became the fashion and a declaration of the post-punk grunge movement. His music and lyrics were poetic anthems for how disgruntled Generation X'ers felt. The blond bohemian singer-songwriter's popularity was massive. His drug habit and unhappiness, just as large. And in the end, his depression, a box filled with heroin, a $500-a-day habit, and a loaded shotgun trumped adoration and fame.

In 1986, the same year Nirvana was formed, Kurt tried heroin for the first time. The band and his drug use spiraled into a blurry, messy mix, and by the 1990s he was addicted to the highs both provided. The following year, at an L7/Butthole Surfers concert, he met Courtney Love, an unstable junkie rocker. Their courtship was short and intense: a year later, the inseparable codependent duo was married and had a child, Frances Bean.

The rest of his life reads like any high-profile rock-and-roll story where names and places can easily be inserted or substituted: Jim Morrison, Jimi Hendrix, Janis Joplin. Each only twenty-seven, each massively famous, each riddled with drug addictions—with one exception: Kurt *meant* to kill himself. The following timeline tracks his *Alice in Wonderland*-like spiral into a huge, and surprisingly documented, black hole.

1993: After injecting himself with heroin in May, Kurt almost ODs. The police arrive at his home to do damage control. Two weeks later, they make another appearance, this time removing four guns from the house. In June, police appear again because of domestic disturbances. He is found high and out of it. Courtney informs the police that it was she who injected him. In July, after another overdose, from which Courtney revives him, he plays at the New Music Seminar in New York City that same night as if nothing happened.

March 4, 1994: After being diagnosed with bronchitis and severe laryngitis, Kurt, who is on tour with Nirvana in Munich, flies to Rome for treatment. He and Courtney check into a hotel. The next morning, she finds him in a coma, having overdosed on champagne and Rohypnol—which her doctor had prescribed for her as a substitute for an antidepressant.

After a five-day stint in the hospital, he's released and returns to Seattle. Many consider this his first suicide attempt.

March 18: Courtney calls the police saying her husband is suicidal and armed with loaded guns. The police find him locked in the bathroom and insisting he's trying to get away from his wife. They remove a number of guns from the house.

March 22: A taxi picks up the couple from their home and drives them to a used-car lot. They fight the entire way. The owner of the '65 Dodge Dart they purchase for $2,500 says that Courtney looks unstable and drops a handful of pills while heading to the restroom.

March 25: Family, friends, and record executives meet for an intervention organized by Courtney in the hopes of getting him to enter rehab. After much arguing, he agrees to a detox program in Los Angeles. Though friends drive him to the airport, he changes his mind and returns home. Courtney flies to LA instead and checks herself into the Peninsula Hotel, where she undergoes outpatient drug treatment.

March 30: Kurt asks his friend to buy him a gun, insisting that prowlers are damaging his house. It's unclear if the pair completes this task together. The owner of the store says two people entered, but one hung back, making it impossible to be identified. Then, for whatever reason, Kurt checks himself into the Exodus Treatment Center. His daughter and her nanny visit him. This is the last time Frances, only nineteen months old, will see her father.

April 1: On his second day at rehab, at 6:00 PM, Kurt tells staffers he's going out for a smoke. He climbs a fence and runs away. Without funds to pay for a flight back to Seattle, he barters an autograph for plane fare for a 10:20 PM flight on Delta.

April 2: He lands in Seattle and goes to his home, bought only four months earlier. A friend who is staying there later comments that he looks terrible. A taxi driver admits to taking Kurt to pick up rifle shells. After learning that her husband has left the hospital, Courtney puts a hold on his credit cards and hires Tom Grant, a private investigator, to find him.

April 8: His body lies on the floor unnoticed in a spare room above their garage for more than three days before Gary Smith, an electrician who comes by to install security lighting, discovers him while peering through the window.

He is dressed in jeans, a long-sleeve shirt, black sneakers, and a tan corduroy coat—his signature outfit. A shotgun rests on his chest, his left hand wrapped around the barrel, which points toward the left side of his chin. Blood is leaking out from his head. The electrician calls 911 and the chaos begins. Fifty minutes later, radio station KXRX releases the news that Kurt is dead. His sister hears the broadcast and phones the station, crying hysterically and accuses them of playing a terrible joke on listeners. TV crews, reporters, police, photographers, friends, and family arrive at Kurt's house. By the end of the day, news of his death is national.

Inside the greenhouse, a cigar box filled with drug paraphernalia—syringes and a spoon—along with a wallet containing $120, cigarette butts, a lighter, a pair of sunglasses, and shotgun shells and their receipt, surround his body. On a nearby table is a handwritten suicide note with a red pen stabbed through it to hold it in place.

When the police leave and Courtney is finally able to have a private, quiet moment, she enters the room, drops to the floor, and dips her hands in her husband's blood. She puts on his blood-splattered coat, which she begged the police to let her keep, and searches the floor for bits of his remains, finding a small piece of his skull—hair still attached—which she promptly washes.

The following day, she's taken to the chapel where Kurt's body is lying, where she says goodbye. The casket is open; he's been cleaned up, blood removed, the traces of his suicide wiped away, his eyes sewn shut. She touches his face, feels his cold cheek, then snips off a lock of his hair. Next she pulls down his pants and clips his pubic hair as a memento.

The hours between April 3 and April 5, when the medical examiner says Kurt died, remain a mystery. Some say he was seen on the third having lunch with a woman in a restaurant in Seattle, that he ran into his manager and bumped into a fan on the street. But these are just rumors, hearsay that can't be proven. They only add to the controversy surrounding his death, which has been the topic of books, documentaries, and investigations. Detectives claim to have spent more than two hundred hours interviewing friends and family. Because many felt that the suicide note was written by someone else, police hired a handwriting expert, who confirmed that the scrawl was Kurt's. During the days leading up to his death, and on the day he died, someone tried using his credit card five different times, yet never in person. The private detective and a friend checked the Cobain home on April 7, but neither bothered to check the spare room. That same day, police arrested Courtney in her vomit- and bloodstained hotel room, where they found a blank prescription pad and what they thought was a bag of heroin. Once released, she checked into a treatment center. She was there only one day when a friend arrived to break the news of his death. But before she could say anything, Courtney instinctively asked, "How?"

Questions still remain. Who was trying to use the credit card Courtney had put a hold on? Was anyone in the room with Kurt before he died? Could Courtney, a woman as messed-up, drugged-up, and temperamental as her husband, have killed him? Heroin and tranquilizers were found in his system, but not enough to kill him. TV shows, magazines, and newspapers all ran stories about conspiracy theories. In the end, investigators, medical examiners, and police officers confirmed that the only one pointing the gun was Kurt.

April 10: The vigil at Viretta Park acts as a grave site, since he was cremated and his ashes scattered. Thousands of fans continue to gather in the park on the anniversary of his death to mourn their icon and remember his contribution to music, making his one of the most celebrated and highly publicized deaths to date.

Kurt's missing person report

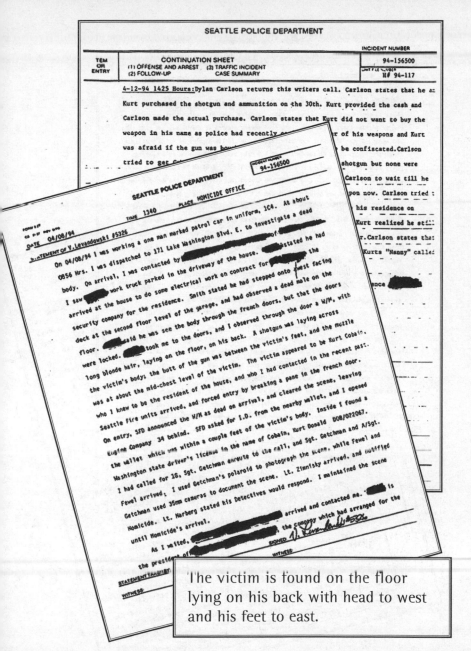

SEATTLE POLICE DEPARTMENT

INCIDENT NUMBER

94-156500

TEM OR ENTRY

CONTINUATION SHEET
(1) OFFENSE AND ARREST   (3) TRAFFIC INCIDENT
(2) FOLLOW-UP   CASE SUMMARY

H# 94-117

4-12-94 1425 Hours: Dylan Carlson returns this writers call. Carlson states that he a...
Kurt purchased the shotgun and ammunition on the 30th. Kurt provided the cash and
Carlson made the actual purchase. Carlson states that Kurt did not want to buy the
weapon in his name as police had recently ... r of his weapons and Kurt
was afraid if the gun was h... be confiscated. Carlson
tried to get ... shotgun but none were
... Carlson to wait till he
... pon now. Carlson tried ...
... his residence on
... Kurt realized he st...
r. Carlson states tha...
Kurts "Nanny" calle...

SEATTLE POLICE DEPARTMENT

94-156500

PLACE HOMICIDE OFFICE

TIME 1340

DATE 04/08/94

STATEMENT OF V. Levandowski #5326

On 04/08/94 I was working a one man marked patrol car in uniform, 1C4. At about
0856 Hrs. I was dispatched to 171 Lake Washington Blvd. E. to investigate a dead
body. On arrival, I was contacted by ... stated he had
I saw ... work truck parked in the driveway of the house. ... the
arrived at the house to do some electrical work on contract for ... west facing
security company for the residence. Smith stated he had stepped onto ... the doors
deck at the second floor level of the garage, and had observed a dead male on the
floor. ... said he was see the body through the french doors, but that the doors
were locked. ... took me to the doors, on his back. A shotgun was laying across
long blonde hair, laying on the floor, on his back. A shotgun was laying across
the victim's body; the butt of the gun was between the victim's feet, and the muzzle
was at about the mid-chest level of the victim. The victim appeared to be Kurt Cobain.
who I knew to be the resident of the house, and who I had contacted in the recent past.
Seattle Fire units arrived, and forced entry by breaking a pane in the french door.
On entry, SFD announced the W/M as dead on arrival, and cleared the scene, leaving
Engine Company 34 behind. SFD asked for I.D. from the nearby wallet, and I opened
the wallet which was within a couple feet of the victim's body. Inside I found a
Washington state driver's license in the name of Cobain, Kurt Donald DOB/022067.
I had called for 1G, Sgt. Getchman enroute to the call, and Sgt. Getchman and A/Sgt.
Fewel arrived; I used Getchman's polaroid to photograph the scene, while Fewel and
Getchman used 35mm cameras to document the scene. Lt. Zimnisky arrived, and notified
Homicide. Lt. Narberg stated his Detectives would respond. I maintained the scene
until Homicide's arrival.

As I waited, ... is
the president of ... arrived and contacted me. ... is
... the company which had arranged for the

SIGNED V. Levandowski #5326

STATEMENT TAKEN BY

WITNESS

WITNESS

The victim is found on the floor
lying on his back with head to west
and his feet to east.

His autopsy report

DATE 04/08/94   TIME 1340   PLACE HOMICIDE OFFICE

STATEMENT OF: V. Levandowski #5326 -continued-

electrical work to be done by ▓▓▓ ▓▓▓ stated he had been informed of the event
by ▓▓▓ and had responded to see if he could be of assistance. ▓▓▓ stated he had
recently spoken with Courtney Love, Cobain's wife, who was in Los Angeles. Love had stated that they were
concerned that there might be unauthorized people staying in the house, and that Love
had arranged for Tim Grant, a private investigator in Los Angeles to go to Seattle to
check the house. Love stated that Grant had been given ▓▓▓ name as a contact.
Pelly said this conversation occurred on 04/06/94. ▓▓▓ stated he received a call
on 04/07/94 at about 0245 from Grant, who stated that Grant was in the driveway of
Cobain's house, and was going to check the interior of the house. At about 1400 hrs.
that same day, ▓▓▓ and ▓▓▓ surveyed the property to assess the wiring job that
▓▓▓ was to perform. ▓▓▓ stated that neither he, nor ▓▓▓ looked into the room
over the garage. Later that day, at about 2140 hrs., ▓▓▓ received another call
from Grant. Grant asked if ▓▓▓ had locked the window that Grant had used to get
inside the house. ▓▓▓ told Grant he had not, as ▓▓▓ had not entered the house.
▓▓▓ stated he had no information on Grant other than his name, and a cellular
phone number of ▓▓▓▓▓▓ ▓▓▓ stated he had never met Grant. I told ▓▓▓
that if he should hear from Grant, to have Grant call the Homicide office.

Once the detectives arrived, A/Sgt. Fewel and I check the interior of the main
house. Nothing appeared to be amiss, and there was nothing of note discovered.

Inside the scene, I had observed a cigar box lying next to the victim. Inside the
box were syringes, a spoon, and other items of narcotics paraphernalia. On a nearby
table was a paper placemat, with a hand-written note in red ink. The pen was stabbed
into the note, holding it in place. The note was apparently written by Cobain to his
wife and daughter, explaining why he had killed himself. I stayed on scene, until
relieved by second watch patrol units, after Cobain's body had been removed by the M.E.

STATEMENT TAKEN BY: self   SIGNED: *V. Levandowski #5326*

WITNESS: WITNESS:

PAGE 2 OF 3

---

of
on the security
ere was a dual
french door
victim was
ss fragments
ce coming to
e area and
e. He has
ence on a
OL from a
had filed
that the
. He also
ody in

...way from ... of Lake Washington
...es of the property. ...e detached garage is located on the
greenhouse above the vehicle are flat. Detectives note that there is a
the westside leading to the detached double garage. All four
french doors on the eastside which lead to the french door entry and another set of
are unlocked and closed but there is a balcony. These doors
gardening supplies on it in front of the door. There is a stool with a box of
the west wall and there are stainless steel planting trays on the
north and south walls. One of the stainless trays contains a pile
of dirt with bulbs in it. On top of this dirt pile is a sink on
written in red ink and stuck into the dirt pile with a red pen.
This is a suicide note directed to Courtney and ▓▓▓ and
signed Kurt Colbain. The victim is found on the floor lying on
his back with his head to the west and feet to the east. There is a

INVESTIGATING OFFICER   SERIAL   UNIT
Det. Jim Yoshida 3168/32

INVESTIGATING OFFICER   SERIAL   UNIT
Det. Steve Kirkland 3356/32

APPROVING OFFICER
[signature]

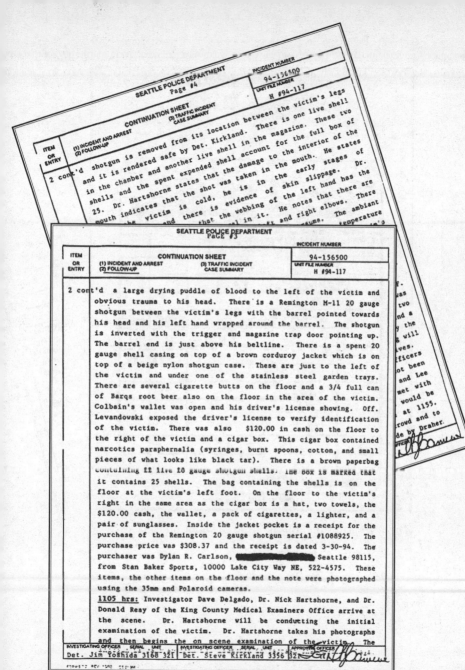

INCIDENT NUMBER
94-156500
UNIT FILE NUMBER
H #94-117

CONTINUATION SHEET
(1) INCIDENT AND ARREST
(2) FOLLOW-UP
(3) TRAFFIC INCIDENT
CASE SUMMARY

ITEM
OR
ENTRY

2 cont'd ...shotgun is removed from its location between the victim's legs and it is rendered safe by Det. Kirkland. There is one live shell in the chamber and another live shell in the magazine. These two shells and the spent expended shell account for the full box of 25. Dr. Hartshorne states that the damage to the interior of the mouth indicates that the shot was taken in the mouth. He states ...he victim is cold, he is in the early stages of ...and there is evidence of skin slippage. Dr. ...that the webbing of the left hand has the ...al in it. ...and right elbows. There ...ume. The ambient ...temperature ...

INCIDENT NUMBER
94-156500
UNIT FILE NUMBER
H #94-117

CONTINUATION SHEET
(1) INCIDENT AND ARREST
(2) FOLLOW-UP
(3) TRAFFIC INCIDENT
CASE SUMMARY

ITEM
OR
ENTRY

2 cont'd a large drying puddle of blood to the left of the victim and obvious trauma to his head. There is a Remington M-11 20 gauge shotgun between the victim's legs with the barrel pointed towards his head and his left hand wrapped around the barrel. The shotgun is inverted with the trigger and magazine trap door pointing up. The barrel end is just above his beltline. There is a spent 20 gauge shell casing on top of a brown corduroy jacket which is on top of a beige nylon shotgun case. These are just to the left of the victim and under one of the stainless steel garden trays. There are several cigarette butts on the floor and a 3/4 full can of Barqs root beer also on the floor in the area of the victim. Colbain's wallet was open and his driver's license showing. Off. Levandowski exposed the driver's license to verify identification of the victim. There was also $120.00 in cash on the floor to the right of the victim and a cigar box. This cigar box contained narcotics paraphernalia (syringes, burnt spoons, cotton, and small pieces of what looks like black tar). There is a brown paperbag containing 22 live 20 gauge shotgun shells. The box is marked that it contains 25 shells. The bag containing the shells is on the floor at the victim's left foot. On the floor to the victim's right in the same area as the cigar box is a hat, two towels, the $120.00 cash, the wallet, a pack of cigarettes, a lighter, and a pair of sunglasses. Inside the jacket pocket is a receipt for the purchase of the Remington 20 gauge shotgun serial #1088925. The purchase price was $308.37 and the receipt is dated 3-30-94. The purchaser was Dylan R. Carlson, ████████████ Seattle 98115, from Stan Baker Sports, 10000 Lake City Way NE, 522-4575. These items, the other items on the floor and the note were photographed using the 35mm and Polaroid cameras.
1105 hrs: Investigator Dave Delgado, Dr. Nick Hartshorne, and Dr. Donald Reay of the King County Medical Examiners Office arrive at the scene. Dr. Hartshorne will be conducting the initial examination of the victim. Dr. Hartshorne takes his photographs and then begins the on scene examination of the victim. The

INVESTIGATING OFFICER SERIAL UNIT   INVESTIGATING OFFICER SERIAL UNIT   APPROVING OFFICER
Det. Jim Yoshida 3168 321   Det. Steve Kirkland 3356 321

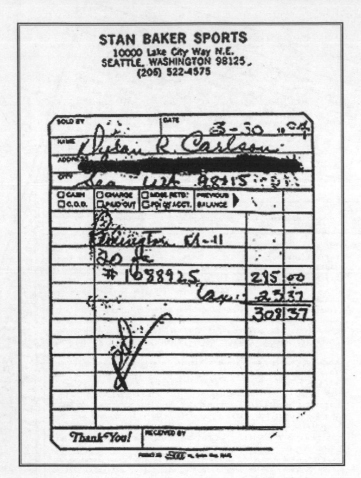

The receipt for the gun he used to shoot himself

UNEARTHED: "Semi-automatics, like a nine-millimeter, are the most popular because they're very easy to get a hold of," explains Dr. Richard L. Keller, M.D., a Lake County coroner, who sees more than sixty suicides a year. "Hunting rifles rank second in popularity." Because of the amount of force a fired gun creates, the bullet isn't always what does the most damage. When fired into the head, a gun pulls two hundred liters of air into the skull, which, acting like a vacuum, can literally blow the brain—sometimes intact—out of the head. "Often parts of the brain and skull are blown backward, landing on the hand and arm of the victim," Dr. Keller adds.

DID YOU KNOW: On March 30, Kurt asked a friend to purchase the Remington 11 twenty-gauge shotgun, serial number 1086925. A month after his suicide, the gun was finally checked for fingerprints. Though four prints were found, none except Kurt's were legible, since by the time the gun was recovered, rigor mortis had set in and the gun had to be pried from his left hand. Courtney later donated the Remington, along with other firearms her husband owned, to Mothers Against Violence.

CAREER HIGHLIGHTS: Considered the most influential punk band since The Sex Pistols, Nirvana transformed music and a generation. In a 2003 Top 100 list, *Rolling Stone* magazine called Kurt the twelfth best guitarist in history. The band recorded only three studio albums, its most successful being *Nevermind*. "Come As You Are," "Smells Like Teen Spirit," and "All Apologies" became some of their most noted songs. At the 1991 Reading Festival, Eugene Kelly of The Vaselines, one of Kurt's biggest influences, joined Nirvana onstage for a duet of their highly regarded "Molly's Lips." He insisted this was one of the greatest moments of his life. (The other was the birth of his daughter.) The band recorded a tribute album in which many famous punk bands and musicians such as Dee Dee Ramone, The Vibrators, and Agent Orange paid homage to the late singer-guitarist.

A number of books and DVDs, including Kurt's own journal, have been released since his death. In 2008, NME.COM reported that Courtney had announced that Ryan Gosling would play her husband and Scarlett Johansson would portray her in a film version of Charles R. Cross's 2002 book, *Heavier Than Heaven: A Biography of Kurt Cobain*, to which Courtney had acquired the rights. In 2005, film director Gus Van Sant based his movie *Last Days* on what might have happened in the final hours of Kurt's life.

SID AND NANCY: Huge fans of punk icon Sex Pistol Sid Vicious and his girlfriend Nancy Spungen, Kurt and Courtney would check into hotels registered under the names of the notorious drugged-out couple. Courtney also lobbied to play the part of Nancy in the 1986 film *Sid and Nancy*. Though she didn't get the lead role, she was cast as a fellow junkie.

The volatile history of Sid and Nancy culminated in their controversial deaths, hers in mid-1978 and his months later. Strung out on heroin, the couple signed into New York's famous Chelsea Hotel in August 1978. Nancy, Sid's twenty-year-old lover turned manager, was found clad in a black bra and panties, sprawled on the bathroom floor, having bled to death from a single stab wound to her stomach by a knife she gave Sid the day before. Asleep several feet away, the rocker awoke unable to recall a thing. He was arrested for murder but released on bail. Ten days after her death, Sid attempted to kill himself by slashing his arms and was taken to Bellevue Hospital. Not long after his release, he was arrested again for assaulting Patti Smith's brother, Todd, when he smashed a beer mug in his face. After spending fifty-five days at Riker's Island, he was bailed out on February 2, 1979.

Clean and sober, he went to his new girlfriend's apartment in Greenwich Village, only to be found dead from a heroin overdose the next day. It's still unclear if his death was intended or accidental. A note that could have been a suicide letter was found in the pocket of his favorite jacket by his mother.

Supposedly, before Sid's mother died in 1996, she admitted to sneaking into the girlfriend's apartment and injecting her son with enough heroin to kill him while he slept, stating that she was fearful he'd be found guilty of killing Nancy and didn't want him to go to prison.

Though the bad-boy rockers never met, Kurt and Sid worked together posthumously. Images of them have been used in a poster campaign for Doc Martens—although the shoe company, after fan uproar over the leaked ads, clarified that the images had not been cleared for release to the general public. The dead musicians also shared "some soul" when the sneaker company Converse released limited-edition boots with a design incorporating writings from Kurt's diary. To mark Converse's anniversary, other late celebrities were used as well: Hunter, Sid, and Joy Division's Ian Curtis.

Elliott Smith performing in Manchester, October 4, 2000

Elliott Smith

BORN: August 6, 1969, Omaha, Nebraska
DIED: October 21, 2003, Echo Park, California
AGE: 34
METHOD: Stabbing
DISCOVERED BY: His girlfriend
FUNERAL: No public burial site or memorial was ever formally announced. Tribute concerts were held in both the United States and England. A ten-thousand-signature petition to make Silver Lake, where Elliott had lived, a memorial park in his honor was submitted, but no word on future plans has been announced.
FINAL RESTING PLACE: Elliott was cremated; what was done with his ashes remains a mystery. With no public burial site ever revealed, fans held a memorial ceremony outside LA's Solutions Audio, where the photo for his *Figure 8* album cover was shot. In July 2006, his high school paid homage to him by creating a memorial.

> *"Depressing" isn't a word I would use to describe my music. But there is some sadness in it—there has to be, so that the happiness in it will matter.*
> IN AN INTERVIEW WITH *ROLLING STONE*, APRIL 2000

He sits on a stool, a guitar in his hand, alone on the stage. His greasy, dark hair seems stuck to his head; his skin is severely pocked; his brown eyes are sunken and hollow. The words no longer come easily. He's not sure where to put his fingers on the strings or which chords come next. He stops and starts, trying to find his way, hoping some of his lyrics will lead him back to the place he knows. The packed crowd shouts out the words like game show contestants eager to win a prize. And Elliott Smith is happy to have the help. He laughs, and as he reaches for his bottle of water, his signature leather wristband catches the light. After a moment or two, he finishes the song. It's a sad sight, and everyone knows it, even Elliott who curses apologetically under his breath, his head twisting from side to side in shame.

What happens to him months later is confusing. Said by friends and colleagues to be drug-free, he still seems not right. Supposedly he is back on his feet, his body clean, his mind sober, thanks to a stay at a Beverly Hills neurotransmitter restoration center, which employs an unorthodox technique for treating addiction. Kurt Cobain went through something similar. The last project Elliott is close to completing is his sixth album, but in October of 2003 all that comes to a crashing halt.

They were fighting in their one-story house on Lemoyne Street in LA's Echo Park, located between a Hispanic blue-collar neighborhood and the basin of Sunset Boulevard. Perhaps the fight was over money. Perhaps it was over the state of their relationship, which ironically had been on an upswing—there was talk of having children. It could have been over drugs and alcohol, which seems the most obvious. No one really knows, since Jennifer hasn't spoken about what happened the morning Elliott died.

Jennifer Chiba, an actor turned art therapist, had met Elliott when each won bit parts in *Southlander*, a low-budget movie made three years before, shortly after he migrated to LA in 1999. Though they dated in the beginning, they were merely friends when Elliott moved in with her fourteen months earlier, after he'd split up with his ex-girlfriend. Fresh out of detox, he had reconnected with Jennifer, and out of loneliness or necessity the two had decided to become housemates. Over the past year, they became romantically involved again.

At some point during their fight, Jennifer locked herself in the bathroom. When she heard Elliott screaming, she thrust the door open and found him standing with his back to her. When he turned around, a kitchen knife was protruding from his chest. He was pale and gasping for air. When she pulled the knife out, he remained standing for a moment, took a few steps, and then collapsed. She called 911 at 12:18 PM, started CPR as the emergency operator directed her, and waited for help. Though Elliott was rushed to the County–USC Medical Center, he died an hour later. When questioned by the police, Jennifer, seated at her kitchen table, first noticed a Post-it note: "I'm sorry—Love Elliott. God forgive me."

Elliott's parents divorced when he was very young, and though his mother remarried, he never got along with his stepfather. Raised in Texas, he moved to Portland, Oregon, at age fourteen, to live with his father. That year, he wrote and recorded his first set of songs, and was introduced to an array of drugs. At Hampshire College he formed the band Heatmiser with a school chum. After two additional members were added, the group soon became a fixture in the underground indie music scene. But the perfectionist in him strove for a finer, more professional sound. This perfectionism would ensure his brilliance and his demise—a Shakespearean character flaw. He didn't like to be told how to adjust his music, nor did he appreciate the democracy a rock band needs to survive. In 1994 some members stopped getting along, and Elliott saw an opportunity to go solo.

His songs were like lullabies—melancholy, honest, and soulful. He was shy and soft-spoken with sad brown eyes. He wore a pensive expression while he sang, but when talking to the audience he would flash a mischievous grin while tossing out something witty. When he smiled, his entire face lit up; the frown lines between his brows disappeared. He had no shtick. No twenty-four-piece band or lavish costumes. It was often just Elliott dressed in a T-shirt, a pair of khakis or jeans, and his guitar.

Almost as if filling the music hole created by Kurt Cobain's death, Elliott became the next must-hear "It" singer-songwriter in the mid-1990s, seamlessly mixing folk and punk into a rich, harmonious blend. And like Kurt, he battled depression, alcoholism, and drug addiction for years. He admitted to friends and in interviews to mutilating and burning himself with cigarettes. He told friends he thought about suicide, that he had attempted it once, stabbing himself with a knife. Friend and producer Larry Crane corroborated this by relating how Elliott had shown him a nasty scar on his chest.

He was bad about keeping in touch with friends, and would sometimes distance himself from people. In the late 1990s, friends tried an intervention, confronting him in Chicago and asking him to go into treatment. He'd tried rehab a few times, but had never completed a full program. At his lowest, he was strung out, depressed, feeding a $1,000-plus daily heroin and crack habit. Haunting bars, he'd abruptly disappear if someone recognized him. Sometimes he'd walk the streets late at night with a blanket over his shoulders, muttering.

In 2002 he performed three times and agreed to the first interview and photo shoot he'd done in months. The year he died, his last gig was at the University of Utah's Redfest, in Salt Lake City. During the performance, he forgot chords and lyrics, and lost his train of thought. A reporter for the online magazine Glorious Noise commented: "It would not surprise me at all if Elliott Smith ends up dead within a year." Sadly, he was right.

Though Elliott had talked often of suicide, and had used a knife to try to kill himself before, he was under a psychiatrist's care and was taking medication for his long-term depression. Toxicology tests confirmed normal prescribed levels of antidepressant and ADHD medications in his system, but otherwise he was drug-free.

Yet many details about the crime scene confused family, friends, and professionals. The kitchen knife Elliott used had been sitting out on a cutting board earlier that morning. The coroner's report cited two single-edged laceration wounds to his chest/heart—which had been repaired at the hospital. There were no hesitation marks, which is highly unusual. There were also small lacerations on the palms of his hands, which appeared to be defensive wounds. Perhaps he got them when Jennifer removed the knife from his chest and he tried to stop her. Most bizarre and unnatural was that the stab wounds were made through his clothing. Usually when someone stabs himself—a highly uncommon method of suicide to begin with—he pulls his clothing away to make sure he's hitting the desired target. Professionals also highlight the fact that Jennifer refused to speak with detectives. No one asked why he had wounds on the palms of his hands, if his fingerprints were on the knife, or if her prints were in keeping with someone having pulled the blade out rather than inserting it. However, Jennifer had no mental illness issues and was not a violent person. Given Elliott's history, his self-destructive behavior, his past attempt, and his suicidal tendencies, the coroner and police confirmed his death as a suicide.

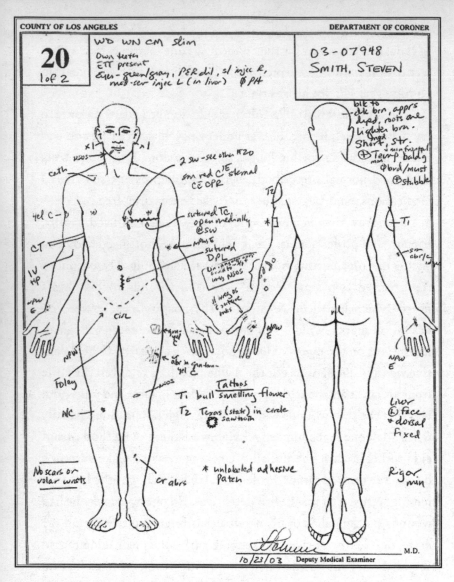

Visual diagram of Elliott's slashed body

UNEARTHED: Self-stabbing in the chest is a rare method, associated with passion and deep personality disturbances. Stab wounds generally reflect the characteristics of the weapon used, the most common being a small steak or carving knife. Other devices include ice picks, forks, broken glass, pens, pencils, pool cues, screwdrivers, and scissors. After a knife has perforated the skin, it easily penetrates tissue and organs with little force needed, unless it meets bone. "Since these knives are so portable, people will carry one around with them and take it out to look at it," explains Lake County coroner Dr. Keller. "Mostly they are trying to bond with the object, making the self-inflicting act less scary."

CAREER HIGHLIGHTS: Elliott was a techno fanatic and music aficionado who had a self-built studio in his home. If he wasn't in front of the mic, he'd be found inhaling energy drinks and fiddling with the soundboard. His first band was called Stranger Than Fiction, which was followed by the just as short lived second group, Heatmiser. *Roman Candle*, his initial solo album, was followed by *Elliott Smith* and then *Either/Or*, which was a huge success. Yet his career took off thanks to film soundtracks, which catapulted him out of his indie niche. In 1997 his song "Miss Misery," one of five of his songs that appeared in the film *Good Will Hunting*, was nominated for an Academy Award. Though he lost to Celine Dion for *Titanic*'s ballad "My Heart Will Go On," his TV performance at the Oscars got him noticed. Wes Anderson used his "Needle in the Hay" in the film *The Royal Tenenbaums*, with the song ironically played during a wrist-slashing scene. The album also featured a dead Nick Drake, and Joey Ramone, who died from lymphoma in 2001. Elliott was also supposed to cover his beloved Beatles' song "Hey Jude" for the *Tenenbaums* soundtrack, but he was in such bad shape he couldn't perform. "Trouble," "Thirteen," and "Let's Get Lost" appear on the soundtrack as well.

From a Basement on the Hill, which was finished and released a year after he died, in October 2004, took him four years to complete. A posthumous two-disc compilation album entitled *New Moon*, which featured twenty-four songs recorded between 1994 and 1997, was released on May 8, 2007.

DID YOU KNOW: After graduation, he changed his name from Steven Paul Smith to Elliott Smith because he considered *Steve* too "jockish" and *Steven* too bookish. When the police filed their investigation notes or when the notes were transcribed to the coroner's report, his name was misspelled as "Elliot."

Michael's coffin being carried out of St. Andrew's Cathedral, Sydney, Australia, November 27, 1997

Michael Hutchence

BORN: January 22, 1960, Sydney, Australia

DIED: November 22, 1997, Sydney, Australia

AGE: 37

METHOD: Hanging

DISCOVERED BY: Housekeeping

FUNERAL: More than a thousand invited mourners filled St. Andrew's Cathedral in Sydney on November 26, and just as many fans flooded the streets. A loudspeaker system and massive TV screens let everyone feel part of the service. Many of Michael's ex-girlfriends attended—Helena Christensen, Kylie Minogue, Michele Bennett—as did Tom Jones, Bono's wife, and Nick Cave, who sang "Into My Arms."

FINAL RESTING PLACE: Michael was cremated at the Glebe Mortuary. His fiancée, Paula Yates, flew back to Britain with one third of his ashes. The remainder was shared by his divorced parents. His father later sprinkled his portion in Sydney Harbor.

It's just as difficult to live in a self-made hell of privacy as it is to live in a self-made hell of publicity.
IN AN INTERVIEW WITH AUSTRALIA'S *WHO* MAGAZINE, 1993

"Never Tear Us Apart" blared mournfully through the loudspeakers of St. Andrew's Cathedral. It was a song that would later become Michael Hutchence's anthem. The casket, adorned with five hundred blue irises and a single yellow tiger lily, symbolizing his sixteen-month-old daughter, was solid and heavy. The members of INXS and Michael's brother were responsible for transporting his body through the chapel and into the hearse. Each was numb from trying to piece together the information surrounding his death while trying to digest the fact that the lead singer of INXS, and one of Australia's most famous personalities, was dead.

At the time of his death, INXS had been performing their twentieth-anniversary Lose Your Head tour for the past few months. The final leg was Australia, a fitting stop that had significance for the band. Having gotten their start there twenty years earlier, they now felt that everything had come full circle.

On the morning of November 22, band members waited in the studio for their front man to arrive for rehearsal. He never showed. Instead he was found dead in his hotel room, naked, kneeling on the floor, his body facing the door. A housekeeper accidentally stumbled upon him when she entered to clean the room.

A black leather belt with a broken buckle lay near his body. The bed was unmade. Clothes were scattered everywhere. Two suitcases lay open on the floor. Film scripts sat on a chair. There were empty beer bottles and unwashed glasses. There were drugs, too: a Becloforte and Ventolin inhaler, painkillers, two hundred tablets of Zovirax, Prozac, and nicotine patches, among others.

In 1977, six school chums were eager to start a band. Originally called The Farriss Brothers, INXS was conceived at Australia's Davidson High School. Within a matter of years, the band went from traveling in a tired-looking van to arriving at venues in limos, from performing in unknown bars to filling concert halls. Of the six band members, Michael stood out, not just because of his deep brooding baritone voice and his ease at creating strong lyrics, but also because of his hunky good looks and fierce personality. He was tall and beefy, with a lion's mane of wild dark hair and piercing, intense brown eyes.

Over the past few years before his death, Michael had struggled against a number of vices. A party boy, he enjoyed his share of drinks and drugs. In 1992, after spending an evening at a Copenhagen nightclub with his girlfriend, he was standing in the street and a taxi attempted to go around him. The driver, who felt he hadn't moved out of the way fast enough, stopped the car, got out, and pushed him so hard he lost his balance and hit his head on the curb. He neglected the injury until headaches became worrisome. Residual effects from the brain damage left him without a sense of smell and only a partial sense of taste. His personality changed, too. He grew depressed and aggressive. While recording *Full Moon Dirty Hearts*, he pulled a knife on a bandmate and threatened to kill him. In an effort to control his depression, he was prescribed Prozac in 1995. Another doctor was brought in two years later when Michael's condition hadn't improved.

The last months of his life were absorbed with getting his fiancée, Paula Yates's ex-husband, Live Aid founder and rock singer Bob Geldof, to let their three children accompany him on the last part of his tour.

Michael and Paula, a television presenter, had met briefly in 1984 when he was a guest on her show. They were both twenty-three, and the attraction was instant and intense. Years later, she interviewed him again, this time, on a bed on the show *Big Breakfast*, her legs draped over his, his body leaning up against hers. The interview was a powerful one, leaving both with deep lust for each other. When she went home, much to her husband's shock, she stuck a photo of Michael on her fridge.

Like the rest of his life, their relationship was a wild ride. He'd dated his share of beautiful women, and often went from one girlfriend to another. After Paula divorced Geldof, she and Michael became inseparable. In 1996 they had a child together, Tiger Lily, his first, her fourth. Uncomfortable being alone with himself and his thoughts, at the time of his death, Michael was looking forward to the tour ending and to finally being able to marry Paula in the beginning of the New Year, in Bora Bora.

On Tuesday, November 18, he checked into the Sydney's Ritz-Carlton hotel under his pseudonym, Murray River, and was given room 524. He was exhausted and frustrated from being on tour, felt invaded by the press, and was mentally drained from helping Paula fight the custody battle for her three children. His priority wasn't seeing old friends, family, or ex-girlfriends; it was getting Geldof to agree to allow his children to visit them in Australia. The day of the twenty-first, he'd been at the recording studio rehearsing. At 7:30 PM, he met his father and stepmother at the Flavour of India Restaurant, where he would spend the next three hours filling them in on his life. Rather than eat, he smoked Marlboro Lights and drank beer. Though he had started out dejected and depressed, his mood quickly switched to jovial. He kissed the restaurant's assistant manager on the lips and danced in the hotel lobby. Either natural or drug-induced, his happiness was evident.

More drinking and drug taking ensued at around 10:45, when his ex-girlfriend Kym Wilson and her boyfriend, Christopher Stollery, stopped by the hotel bar to catch up. The threesome migrated to Michael's top-floor harbor-view room, where more cocktails, vodka, beer, champagne, and coke were consumed.

Concerned about the phone conversation he was scheduled to have with Paula's ex-husband, he asked the pair to stay. They talked about old times, about the film career Michael was hoping to start, about his disappointment about being overlooked for a part in Quentin Tarantino's film *From Dusk Till Dawn*, about the favorable things Miramax had said about him. He made calls throughout the evening, with the couple going out on the balcony from time to time to give him privacy, but as night blurred into gray, and gray revealed blue sky, his friends finally said a tired goodbye close to 5:00 AM. Before they left, Michael crawled into bed and mentioned how he'd love to have a Valium.

Then the phone calls started. The first was to Paula's ex-husband, who told Michael the courts had been adjourned until December

17, which meant that he would be finishing the tour alone. Michael became angry, the volume of his voice rising. It was loud enough to wake the woman in the next room, who recalled hearing shouting and was able to make out bits and pieces, most clearly, "She's not your wife anymore!"

Paula phoned at 5:31 AM to reiterate what he already knew.

Enraged, he called Geldof again at 5:38 AM and begged him for the kids, but Geldof refused to let them miss their last three weeks of school. The call ended at 5:54 AM, when Michael slammed down the phone.

At 6:09 AM, he reached out to another ex-girlfriend, Michele Bennett, but got her machine. He left a drunken, somewhat upset, and irrational message, a type she'd grown accustomed to receiving. More calls followed, several to his personal manager telling her he'd had enough and wanted out. Another was to Michele again, at 9:54 AM, where he cried, sounding more depressed than before. Then he rang the front desk and, using his phony name, left a message for his tour manager: "Mr. River is not going to rehearsals today."

Panicked, Michele arrived at the hotel forty minutes later. She knocked on his door and, after receiving no answer, assumed he'd gone to sleep. She then used the house phone to call his room, but gave up after four rings. She left a note for him with the front desk saying she'd wait for him at a nearby café.

According to the coroner's report, a knot from Michael's belt had been tied to the automatic door closure. He had strained his head forward into the loop so hard that the buckle broke. After management was alerted, followed by the police and medical examiners, he was pronounced dead at 12:30 PM.

Some rumors dismissed the death as a suicide, suggesting that an appreciation for rough sex had led Michael to an act of autoeroticism that involved self-strangulation. Gossip took him from rock star to tabloid celebrity in a matter of days, and though dismissed by the police and coroner, the autoeroticism rumor stuck, making his suicide more memorable than his body of work.

With the pack of people surrounding him—personal manager, accountant, lawyers, tour manager, bodyguard, publicist, record company, management company, roadies, assistants, bandmates, friends and family, and his fiancée—it's impossible to imagine that between 9:50 and 11:00 AM there wasn't anyone he could have called to talk him down.

Though "Never Tear Us Apart" became his anthem, other of his songs—"The Devil Inside," "Slide Away," or "Need You Tonight"—seem more fitting to describe him, his life, and his final state of mind.

UNEARTHED: It's estimated that between five hundred and one thousand deaths from autoerotic asphyxiation, strangling or suffocating oneself to heighten sexual arousal and orgasm, occur each year in the United States. "Gaspers," as its practitioners are called, purposely cut off their blood supply to the brain. The lack of blood flow and oxygen creates giddiness, lightheadedness, and exhilaration that can heighten the orgasmic experience. Said to be one of the riskiest of all sexual behaviors, autoerotic asphyxiation can cause people to die accidentally from brain damage, suffocation, or strangulation. Hanging and suffocation with a plastic bag over the head are two of the most commonly used methods. In 1791, composer Frantisek

Kotzwara's was one of the first recorded deaths from this odd sexual perversion.

CAREER HIGHLIGHTS: The band changed its name to INXS before releasing their first album in 1980, which featured their first Australian hit single, "Just Keep Walking." Their sixth album, 1987's *Kick*, was a ten-million-copy smash. Follow-up recordings never made the same impact, and 1997's *Elegantly Wasted* was an expensive flop. Michael's long-awaited eponymous solo album was released in 1999; he'd been working on it for more than four years and continued to do so up until three days prior to his death. "Possibilities" was the last song he recorded. By 2005, reality TV stepped in with *Rock Star: INXS*, a program in which the remaining band members searched nationally for Michael's replacement.

PAULA'S LAST DAYS: Paula was a British writer and TV personality, loved most for her work on *The Tube* and *The Big Breakfast*. A rock-and-roller who lived life hard and wild, taking her share of drugs, she had tried to kill herself months before Michael's suicide. Unable to deal with divorce from Geldof and the nasty custody battle, she overdosed on pills, which she washed down with liquor. Over the next few months, she pulled herself together, relying on her upcoming wedding to push her out of the darkness. His death was more than she could take, and she hit a downward spiral. Drug addiction, a stint in a mental institution for depression, another suicide attempt where she tried to hang herself, followed by a rehab visit, were all part of the next few years. While still battling for custody of her children with her ex, and for Tiger Lily with Michael's parents, she grew indigent. Unable to pay her bills and in deep financial straits, she grew more depressed. In 2000, the body of the forty-one-year-old Paula was found naked in her Notting Hill, West London, flat by her four-year-old daughter after she'd OD'd from a heroin binge. An empty vodka bottle and a half-empty bottle of barbiturates were found by her side.

DID YOU KNOW: Before Michael's casket was carried to the hearse, his mother removed the purple-and-yellow Versace handkerchief from his coat pocket and snipped the six Vivienne Westwood buttons from his jacket. After the funeral, the buttons were handed out to family members as keepsakes.

Musicians Unearthed

LEADER OF THE PACT: Exhibitionist and controversial punk rocker Darby Crash, lead singer of The Germs, was only twenty-two when he killed himself in 1980. Known for their poor behavior and controversial riot-causing shows, the group was banned from performing at a number of clubs, yet they still enjoyed a rare popularity throughout the LA punk scene in the late seventies. Because The Germs were novices and could barely play their instruments, Darby would vomit, smear peanut butter on his body, or gash himself with broken glass in the hopes of distracting his fans from his poor playing. Favoring heroin, he intentionally overdosed, having made a suicide pact with his close friend Casey Cola—who ended up surviving the attempt. During his last minutes, Crash tried to write, "Here lies Darby Crash" on a wall, but did not finish. His suicide (which ordinarily would have brought notorious immortality) was overshadowed by the assassination of John Lennon, which happened the following day. Darby's notoriety was resurrected, however, with the biopic *What We Do Is Secret*.

ANOTHER FAMOUS PROMISE: Married since 1984, rock star Ozzy Osbourne and wife, Sharon, have a suicide pact. Should one go first, by whatever means, the other has sworn to swallow a vat of poison. But the agreement doesn't go into effect until 2012.

WENDY O. WILLIAMS: Known for her Mohawk haircut, Wendy O. Williams—aka "The Queen of Shock Rock"—was the lead singer of The

Plasmatics, an underground cult punk band. Like The Germs, The Plasmatics performed such onstage antics as smashing a television with a sledgehammer, chainsawing guitars, and simulating sex wearing only shaving cream. In 1985 the ex-topless dancer and high-school dropout was nominated for a Grammy in the Best Female Rock Vocal category. One of the most controversial and radical women singers, Wendy became a vegetarian and traded in her bad-girl behavior for an animal-loving image. In 1991 she moved to Connecticut to live with her former manager while working as an animal rehabilitator. On a Monday night in April 1998, her boyfriend came home to find a package she had left for him filled with his favorite pasta, seeds to be planted in their garden, massage balm, sealed letters, and a "living will" requesting that she not be placed on life support. Panicked, he searched the woods for more than an hour, finally finding her near their home with a gunshot wound to her head. She was forty-eight. The last line of her suicide note read, "[F]or me much of the world makes no sense, but my feelings about what I am doing ring loud and clear to an inner ear and a place where there is no self, only calm."

OTHER NOTABLE SUICIDES BY WOMEN: Susannah McCorkle, a sultry, kittenish pop-voiced jazz singer jumped off the balcony of her sixteenth-floor Manhattan apartment; and Phyllis Hyman, a bipolar alcoholic singer and actress, killed herself with pills—phentobarbital and secobarbital—just hours before she was to perform at the Apollo.

THE SADDEST SONG: "Gloomy Sunday," aka the "Hungarian Suicide Song," was written in 1933 by Hungarian-born Rezsö Seress, a self-taught pianist and composer. With words by his poet friend Ladislas Javor, it quickly earned a reputation as a "suicide song." Reports from Hungary claimed that people had taken their lives after listening to the haunting melody and depressing lyrics, and thus Hungarian officials prohibited it from being played. Many American and European radio services echoed similar sentiments, and the song was banned from the airwaves. Billie Holiday resurrected it in 1941, making it popular once again. In the song, the singer mourns the untimely death of a lover and contemplates suicide. The song's composer, Seress, a Holocaust survivor, committed suicide in 1968 by jumping out his Budapest apartment window.

English translation:

> It is autumn and the leaves are falling
> All love has died on earth
> The wind is weeping with sorrowful tears
> My heart will never hope for a new spring again
> My tears and my sorrows are all in vain
> People are heartless, greedy and wicked . . . Love has died!
> The world has come to its end, hope has ceased to have a meaning
> Cities are being wiped out, shrapnel is making music
> Meadows are colored red with human blood
> There are dead people on the streets everywhere
> I will say another quiet prayer: People are sinners, Lord, they make mistakes . . .
> The world has ended!

Six
Artists

What's the fastest way to make money on the artwork you own? Hope the artist dies. This isn't a revelation or a secret. It's a bad joke made by everyone from the aficionado collector to the amateur. In the art world, death has a large price tag. The value of a work doubles or triples instantly when its maker dies. People rush to gobble up as much of the remaining art as possible. We know that van Gogh sold only one painting while living. Yet after he killed himself, he skyrocketed to iconic immortality. One of his more honored paintings, *Portrait of Dr. Gachet*, completed the month before he killed himself, fetched a record-breaking price of $82.5 million in 1990. Van Gogh's name appears six more times on the list of most expensive paintings—more than any other artist, sharing that honor with Jackson Pollock and Andy Warhol, each of whose deaths created enormous speculation and controversy. Mark Rothko is there, too, with *White Center*, which sold in 2007 for $72.8 million. Diane Arbus's photograph *Iden-*

tical Twins is tenth on the list of the most expensive photographs, having sold in 2004 for $478,400. These works have only increased in value, and other pieces by these artists remain extremely sought after.

Artists, especially visual ones, have been perceived as struggling, tortured souls as far back as 1584, when Daswanth, an Indian miniaturist painter, stabbed himself with a dagger. Another suicidal artist was Francesco Borromini, a prominent and influential Italian architect, who in the summer of 1667 suffered from severe sadness. He spent weeks without leaving his home in Rome, and one night decided to burn all of his drawings. Weeks later he found a ceremonial sword and fell on it. He survived for twenty-four hours, long enough to write a will.

The reasons artists kill themselves are varied. For the ultrasensitive, rejection can ignite self-destructive tendencies and break one's spirit. Léon Bonvin, a French watercolorist from the mid-1800s, hanged himself from a tree in the forest of Meudon after a Parisian dealer rejected his paintings. Several decades later, Jules Pascin, a Bulgarian-born painter, draftsman, and printmaker, hanged himself in his Paris studio on the eve of an important show, reportedly from sadness due to poor reviews. Plagued by exhaustion, insomnia, and depression, Nicolas de Staël, a Russian abstract landscape painter, leaped to his death from his eleventh-story studio terrace after a disappointing meeting with a disparaging art critic.

Then there's the loss of a loved one. Dora Carrington was an English painter and decorative artist who tried to kill herself, first with carbon monoxide poisoning and then, months later, by shooting herself, after her longtime companion and best friend, Lytton

Strachey, died of stomach cancer. French painter Jeanne Hébuterne was pregnant with her second child when she leaped from a fifth-story window two days after her partner died of tuberculosis. Louis-Léopold Robert killed himself in Venice, in front of his easel, on the tenth anniversary of his brother's suicide. And Constance Mayer slit her throat with the razor of her lover and teacher, painter Pierre-Paul Prud'hon, who had decided not to marry her.

The "broken heart" category easily slips into the "guilt over an affair" category. Austrian painter and draftsman Richard Gerstl, known for his expressive, psychologically insightful portraits, killed himself in 1908 after a brief romantic fling with the wife of composer Arnold Schoenberg. After setting his studio on fire and burning his letters, biographical material, and artwork, he hanged himself in front of his studio mirror and disemboweled himself with a butcher knife. Unfortunately, many of his works were destroyed in the fire.

The cruelty of life is another motivation. Severe depression from a succession of unhappy events led forty-four-year-old Arshile Gorky, an Armenian-born abstract expressionist, to hang himself in 1948. His studio had burned down, his wife had left him, taking their children with her, he'd been stricken with colon cancer, a car accident had left him with a broken neck and temporary paralysis in his painting arm, and he was penniless. Unable to see any kind of light at the end of a very dark and unending tunnel, he saw death as the only illuminating solution.

Whereas actors and musicians live onstage or on the big screen, visual artists, most of whose artistic endeavors aren't discovered or commercialized until they're deceased, work internally and alone.

"Artists are mostly right-brain thinkers, open and imaginative," says Daniela E. Schreier, Psy.D., a forensic and licensed clinical psychologist. "[T]hey are removed from their fan base because these are people who don't necessarily search for the limelight, but rather they see the world a little bit differently and have a need to express that world through their art."

According to Dr. Schreier, visual artists are hypersensitive, ultra-emotional, and usually suffer from social anxiety, often feeling isolated, uncomfortable, and misunderstood. "This is what contributes to being depressed and taking their own lives—high levels of neuroticism. They have a low emotional baseline for experiencing unpleasant emotions: anger, anxiety, depression. They feel these deeper and sooner. Yet, they are avant-garde and ahead of their time," Schreier explains. "They are visionaries who see deeper and live inside themselves, while constantly battling with the world, which they feel is rejecting them and their work. They have a hard time living in this world because they don't fit in."

Art transcends language and cultural barriers, telling us something about the private world and ideas of its creator. Through their art, artists express their most intimate feelings, hopes, dreams, and despairs. Their photographs, paintings, and sculpture all tell a story. This is certainly true for Vincent van Gogh, Diane Arbus, and Mark Rothko. Whereas a song or a film is usually tied to a particular year or culture, art still fascinates and retains its meaning and value for centuries.

The three artists highlighted here each cut themselves—in one way or another, a more violent method, according to some specialists, since pain is involved. Their self-mutilating act was the exact opposite of the quick result brought by a single shot to the head, though van Gogh did both. (He cut off his ear with a razor, and years later shot himself.) And let's not forget the visual aspect. Cutting, slitting, stabbing is messy, bloody, and visually shocking.

The Café de la Mairie, now named the Vincent van Gogh Café, where the artist lived and died, Saint Remy, France, 1955

Vincent van Gogh

BORN: March 30, 1853, southern Netherlands
DIED: July 29, 1890, Auvers, France
AGE: 37
METHOD: Gunshot in the lower chest/upper stomach
DISCOVERED BY: Owner of the inn
FUNERAL: On July 30, at 3:00 PM, good friends Père Tanguy, Emile

Bernard, and Lucien Pissarro, along with family, attended.

FINAL RESTING PLACE: Interred at the Auvers-sur-Oise cemetery. His headstone reads "Ici repose Vincent van Gogh 1853–1890" ("Here Lies Vincent van Gogh"). Six months later, his brother Theo joined him.

The sadness will last forever.

SOME OF HIS LAST WORDS, SPOKEN TO HIS BROTHER THEO

Paris is dark and loud. Inebriated tourists mesh with locals as both trickle out of bars and spill onto the streets. A man wearing a hat rigged with lit candles sits on a wooden stool on the sidewalk, painting. At his side is an easel; a wooden pipe sticks out of his mouth. His clothing is shoddy, his red beard unkempt. Those who get too close or ask him questions are quickly dismissed, shooed away like annoying flies.

Before Vincent van Gogh is born, his mother gives birth to a boy who is stillborn. The eldest of six, van Gogh is named after the dead child, which many feel contributes to his emotional instability. As a youngster he's silent, clumsy, and moody. He is always thought of as eccentric, with quirky mannerisms, and over the thirty-seven years of his life his behavior grows frighteningly manic and disruptive.

Early on he has trouble holding down jobs. At sixteen he apprentices with an art dealer, during which his passion for art develops. Soon fired, he tries his hand at teaching. When that fails, he gravitates to the ministry, quickly becoming fanatical about religion—a pattern he returns to many times throughout his life. That, too, ends badly. Then, after losing a job in a book shop, he finally turns to art.

His mother is chronically depressed, so van Gogh, aching for love, turns his need to be nurtured into a number of unfortunate affairs. Sent into a deep depression over the rejection of his first love, the daughter of his London landlady, he turns his attention to his cousin. To demonstrate to her how great are his feelings for her, he holds his hand over a lantern's flame. When his father finds out, he is booted from his home. With few options, he moves in with a pregnant prostitute, bringing further shame to his family.

The only person he keeps in constant contact with is his younger brother Theo—a codependent relationship in which the two are like twins.

After spending six years dabbling as an artist—withdrawing from art school in the Netherlands and Belgium—he joins Theo in Paris in 1886. His Dutch upbringing is swapped for a French lifestyle, as he replaces his homeless attire for a suit and a colorful scarf. He hangs out with other artists, becoming strongly influenced by the Impressionist movement, drinking and smoking while obsessing over perfecting his craft. Within two years he completes twenty-five self-portraits. But he grows moody, lashing out in anger. He stops taking care of himself. His teeth fall out; he becomes skeletal, as if he is purposely starving himself. He refuses to leave the house. When guests come for dinner, he eats alone in a corner. He becomes a religious fanatic again. At his worst, he confesses to beating himself or locking himself out of his house and sleeping in a cold shed without a blanket as penance for his failures.

In 1888 he moves to Arles with the intention of starting an artist's colony. He asks his friend and fellow artist Paul Gauguin to move in and help him fulfill his vision. Squeezing tubes of paint directly onto the canvas, he develops a style that is spontaneous and instinctive. He paints self-portraits and portraits of friends, blossoming fruit trees, sunflowers, landscapes, and houses—all of which mark his first great period.

While his work excels, his fighting with Gauguin grows unbearable, until the latter threatens to leave. In a counterattack, van Gogh throws a glass in his face. Another time he pursues him with a razor. On Christmas Eve, van Gogh cuts off the lower part of his left ear, wraps the bloody piece in newspaper, and gives it for safekeeping to his favorite prostitute at the local brothel. She alerts the police. When they arrive, they find the artist unconscious, and they take him to the hospital, where he falls into a psychotic state requiring three days of solitary confinement.

Additional attacks follow, requiring more hospital stays. During this time, Theo marries, has a child, and becomes financially stable, the polar opposite of his brother, who is single, degenerate, and living off the money Theo has been sending him. Plagued with illness—hallucinations, possible seizures, and depression—van Gogh attempts suicide several times by drinking turpentine and swallowing paint and other poisons. Without options, he enters St. Rémy, a mental institution, where he stays for a year. While there, he produces more than three hundred works. Most noted is *Starry Night*.

Released in 1890, he returns to Paris to visit his brother, sister-in-law, and their new child, hoping the vibrancy and stimulation there will inspire him. But the city proves too wild for his new mental-ward mentality, and he resumes drinking and smoking, until he's introduced to a new doctor, Paul Gachet, an understanding artist who tends to the creative. With the goal of calming his nerves and being closer to his new doctor, van Gogh moves to Auvers, just north of Paris, where he spends the last ten weeks of his life.

With Theo's money continuing to pay for his lifestyle, he rents an apartment above a café. He and Dr. Gachet develop an unhealthy relationship, blurring the patient-doctor boundary. Yet something must be working, because van Gogh's mood improves: He stops drinking and having seizures. Most of his mornings are spent painting outdoors, and the afternoons, working in the back room of the café. He paints portraits, first of the doctor, then of his twenty-year-old daughter, for whom he develops strong romantic feelings. During this time he finishes seventy paintings and thirty drawings, all of which burst with color and a feeling of calm—cottages, flowers, sunny landscapes. His paintings are cooler; his work more expressive, less structured and controlled. As with many who are suicidal, or are living in a chronic state of depression, van Gogh seems to experience creative clarity and a prolific period.

In early July, just a month or so later, everything falls apart. He receives a letter from Theo stating that his son is ill and they can't send him any more money. Already a burden to his brother and his family, van Gogh feels abandoned and in competition with his ailing nephew. He also believes all the terrible things that are happening to Theo and his family are his fault. Even though he visits and sees for himself how exhausted the family is, he still begs for the 150 francs he's been receiving monthly for paint, supplies, and living expenses. Theo complies.

Van Gogh starts doubting his doctor's competency and writes to his brother that Gachet is sicker than he. "When the blind lead the blind, don't they both fall in the ditch?" On Sunday, July 27, only several days after returning from Paris, van Gogh shows up at Dr.

Gachet's house and, after seeing that a painting he had admired, of a woman lying on a couch, has not yet been framed, twists into a massive rage. The two men argue, and though van Gogh has a gun in his pocket, rather than stick it in the doctor's face, he leaves.

In the late afternoon of that same day, van Gogh walks into the countryside, sets up his easel, and paints in the fields against a haystack. He removes the revolver he's been carrying around with him for the past several days—having said he needed to borrow it to shoot irritating crows—aims it at his chest, and pulls the trigger. The bullet passes just under his heart. Unsure of what has transpired, he tries to get up several times, falling each time, and then somehow stumbles home.

Having lost a tremendous amount of blood, a bent-over and pale van Gogh enters the inn at around dusk. The owner asks if anything has happened, since the painter is returning late and looks sickly, but he insists that he's fine. Not believing him, the innkeeper goes upstairs and finds him in bed with his face to the wall, his body in a fetal position. Van Gogh shows him his wound, uttering, "I wanted to kill myself."

Dr. Gachet is the first to arrive, at around nine o'clock that evening. He examines the wound, tries to calm van Gogh by lying that it doesn't look bad, and asks for Theo's address. Van Gogh refuses. Gachet obtains the address anyway and writes Theo a note. Arriving in the morning, Theo rushes upstairs, gets into bed with his brother, and cradles him. It's surmised that the bullet is lodged in his chest, having been deflected by the fifth rib. As Dr. Gachet is unable to remove it, he tells him to rest. The following day the artist sits up, smokes a pipe, and seems content, comfortable. A rotation of friends and doctors takes turn caring for him. When the police arrive to investigate, he tells them to leave. That evening he grows weak from an infection, and by 1:30 AM, on July 29, 1890, he dies in his brother's arms. His last words are "I would like to go like this." And so he does.

Three days before his suicide he writes two letters to Theo. One is mailed; Theo finds the other, bloody, in his brother's jacket pocket after he dies.

Though Theo prints funeral announcements with the intention of holding a memorial the following day, a suicide is considered an illegal act, and the church won't permit the service. Instead, friends at the inn encircle van Gogh's closed wooden coffin in the room where he used to paint. His coffin is covered with a white cloth, topped with sunflowers and dahlias. His easel, folding stool, and brushes are placed nearby. The group follows the hearse to the cemetery. They walk up the hill, commenting on the beauty of the area and the irony that he will be looking over the objects he painted: the sky, cottages, barns, wheat fields, and flowers.

Van Gogh was a man full of frustrated wants: for acceptance, notoriety, money, good health, a place to live that wasn't too stimulating, a wife and a child like his brother had. His needs for these things grew so great that being around others who had them became intolerable for him. And he never obtained any of them. He died unfulfilled, lonely, angry, and penniless. Completely self-taught and considered one of the greatest Dutch painters of all time, he embodied the image of the tortured artist, the struggling genius whose work and style went unappreciated until he was dead.

Vincent van Gogh and his brother Theo's headstones side by side

---

UNEARTHED: Van Gogh syndrome is when a person attempts self-mutilation while hallucinating. Also called self-injury syndrome, VGS can be common in people with bipolar disorder, borderline personality disorder, or certain forms of Tourette's syndrome.

A FAMILY CONDITION: As with Virginia and Hemingway, van Gogh came from a family riddled with forms of depression. Two of his uncles were diagnosed with mental illness; his mother was depressed; his sister was sent to an asylum toward the end of her life; and one of his brothers killed himself a decade after van Gogh died. Theo also died in a mental institution six months after the loss of his brother. Van Gogh's mysterious symptoms and illness have long baffled medical, psychiatric, and educational professionals. More than 150 psychiatrists have tried to pinpoint his condition, and more than 30 different diagnoses have been suggested—syphilis, poisoning from sucking on brushes dipped in lead-based paint, temporal lobe epilepsy, and negative effects from his massive drinking. Schizophrenia and bipolar disorder remain the top two. To date, no diagnosis has been universally accepted.

CAREER HIGHLIGHTS: Van Gogh started with watercolors while at school, and worked for a year on single figures in black and white, for which he was criticized. He graduated to multiple figures, but destroyed these when Theo thought they lacked liveliness and freshness—and thus he turned to oil paintings, many of which he destroyed as well. *Old Tower in the Fields, The Potato Eaters*, and *The Cottage* are the only ones remaining from that period.

Even among the cheerier works he produced, there was always the feeling that something darker lurked in the shadows, ready to pop out from behind a barn or rise from the wheat fields. The works were painfully honest and highly emotional, attributes of expressionism and something that nineteenth-century European society was not ready for.

Virtually unknown when he committed suicide, he was the subject of only one article during his lifetime; he sold only one painting while living, *The Red Vineyard*, for four hundred francs, just four months before he shot himself. Though Theo said there was a buyer for the painting, many feel he lied and purchased it himself. Van Gogh exhibited almost no work while living, and his first one-man show happened in 1892, almost two years after he'd died. His lack of success and the initial negative reaction from the press derailed his achievements, adding to his masochistic tendencies and constant physical setbacks.

In only eight years he produced more than two thousand works: oil paintings, drawings, and sketches. His best-known pieces were created in the final two years of his life. Some say his *Crows over the Wheat Field* is his last painting. Others think it was *Field with Stacks of Wheat*. Aside from his paintings, his correspondence—one thousand letters to family and friends—especially that between him and Theo, are the best tangible documents that reveal his state of mind.

Diane Arbus

BORN: March 14, 1923, New York City
DIED: July 26, 1971, New York City
AGE: 48
METHOD: A combined overdose of barbiturates and slitting her wrists
DISCOVERED BY: Her friend
FUNERAL: For such a luminary, Diane's memorial at Campbell's Funeral Home in Manhattan was not well attended. Aside from her immediate family, only her ex-husband, Allan Arbus, and his second wife, Bea Feitler, and Ruth Ansel, Richard Avedon, and Frederick Eberstadt came to pay their respects. Many of her friends were away over the summer, and not a lot of people were informed of her death. The *New York Times* didn't print a death notice until August 1. *Newsweek* and *The Village Voice* didn't run theirs until ten days after her passing.
FINAL RESTING PLACE: She was cremated, yet where her ashes are buried remains a mystery.

> *Something is ironic in the world and it has to do with the fact that what you intend never comes out like you intend it.*
> FROM *DIANE ARBUS: AN APERTURE MONOGRAPH*

In the photograph, the air in the seedy, rundown hotel room appears warm and sticky. The wallpaper in the background displays squiggly etchings. In the unmade bed sits a Mexican dwarf, shirtless and naked from the waist down. A towel covers his private parts and much of his right leg—except for a small part of his foot, which peeks

Diane posing for a rare portrait in Manhattan's Automat restaurant
at Sixth Avenue, between 41st and 42nd Street, 1968

out from underneath the cheap fabric. He's leaning on a varnished wooden nightstand, where a bottle of liquor and a dirty glass sit. His fingers are stumpy, his lips are full, his mustache thin. He has a wide nose and thick brows. His expression is almost smug, calm.

Snap.

Click.

A moment is captured.

The small man fills the picture. Tension and intimacy share the space. There's something disturbing and inviting. Unattractive and interesting.

A shy woman with cropped brown hair stands a foot or so away, a camera held at chest level.

Midgets, circus freaks, nudists, retarded children, giants, twins, and transvestites. Human oddities. These were the subjects and the passions behind the controversial black-and-white photographs that made Diane Arbus famous. She made the ugly beautiful. The odd, attractive. She showed freaks as human. Welcomed their unnaturalness, their dark, eerie creepiness, and they in return welcomed her into their lives. They let her follow them from their day jobs and into their homes, where she captured who they were when the crowds went away, when the lights dimmed, and when the rest of the world wasn't looking. She was like the Pied Piper, but rather than toting a flute, she flashed a camera. Set up lights. They went willingly. Seduced by her charm, her big green eyes, her almost boyish body, exotic face, and unthreatening manner, they opened up emotionally.

There were trips to bars in the Bowery, visits to Coney Island, jaunts to a nudist colony in Pennsylvania, outings to the circus. Underground sex rooms where dominatrices worked on clients. Wherever there was controversy, odd events, or sexual happenings, Diane Arbus was there to document them.

She didn't look away, but rather looked into. Forced us out of our comfort zone and into seeing the ugly, the distorted, the odd, and the unattractive. She showed her own unattractiveness, too: which happened in the summer of 1971. For two days she lay dead in her Greenwich Village apartment. Dressed in pants and a shirt, her body was found resting on its side in an empty bathtub, covered in blood from the methodical slits she'd made in her wrists. The barbiturates she'd overdosed on had shut down her system as blood leaked from her wrists in twin streams.

One of three children, Diane (pronounced Dee-ANN) got her name from a character in the play *Seventh Heaven*. Born to wealthy Jewish parents who were part of Manhattan's upper crust, the young Diane resided at high-end zip codes on Central Park and Park Avenue. She often described her parents as distant and cold.

At fourteen she fell in love with Allan Arbus, an advertising employee who worked for her parents, and against their wishes, she married him four years later. Over the next several years the couple had two daughters, Doon and Amy.

The twosome established a successful venture in fashion photography, first by getting work from Diane's parents, then from glossy magazines. Allan took the photos; she styled the models and directed the shots. It was Allan who gave Diane her first camera, and though they shared equal billing on the photos, it was only a matter of time before she wanted to do more than style shots. She wanted to shoot them. So she quit. In the late 1950s, while her marriage fell apart, she took classes with Lisette Model, a famous documentary photographer, who became her mentor and the mother she felt she never had.

In 1959 the couple divorced and she moved to the West Village, where she met her next mentor, and her lover, advertising director Marvin Israel, also a Jewish New Yorker. She started working as a fashion photographer for various advertising agencies and high-end magazines, such as *Glamour* and *Vogue*. Her first assignment for *Esquire* was to take morgue photos at Bellevue. Surprisingly, Allan remained a true force in her life—he developed her film, let her test-drive new cameras.

Diane's work grew darker as her interest in those left of center became more intense. These subjects were hard to look at, harder still to shy away from. Though she never took a photo of someone who didn't want her to, she did have a few tricks to get the desired shot. She would act dumb or drop things, purposely trying to distract her subjects so they'd feel less self-conscious. When she was stuck or had lost her focus, she'd momentarily look away, hoping to reconnect with the reality of making the photograph.

Soon her name received recognition, and her edgy photos became famous. From the 1960s to the '70s she amassed an enormous portfolio. In 1963 she received a Guggenheim Fellowship. Three years later she won another. Her first staged museum show, *New Documents*, opened in 1967 at the Museum of Modern Art.

When she wasn't researching her next shoot, teaching, or spending time with her children, she would gravitate toward the phone, her lifeline. Her need to connect to whomever—her brother or friends, or possible photography connections, police officers, or a morgue assistant—became insatiable.

In 1969–70 she worked on a series of photos of mentally retarded children, a project that enthralled and delighted her. But in June, she changed her mind and decided she hated the photos. She was still teaching but felt unfulfilled, misunderstood, and depressed. It was no secret she had mood swings and fell into funks. Summers were especially hard. They reinforced and exasperated her feelings of abandonment by her parents as a child, and by the housekeeper whom she had loved but who had abruptly left her. Her friends were away, vacationing with their own families, and her daughters were old enough to be on their own. Allan, now remarried, had moved to California. She turned down a position to teach photography at Yale, but agreed to exhibit her work at the Venice Biennale for the upcoming summer, an unprecedented invitation no photographer had received before. But whenever the subject of either opportunity came up, she would break down in tears.

In July, still extremely depressed, she visited friends in the Hamptons. And though they recognized that she was in a dark place, they didn't know how to help. Others noticed she was tying things up. She wrote to friends asking if they wanted letters back that they had written to her over the years. She phoned others to say her children were well settled.

"We had our own lives to lead," admitted a neighbor, quoted in Patricia Bosworth's biography of Diane Arbus. "We didn't know how to deal with it—she was getting so tough to handle—really heavy. It was excruciatingly depressing being with her. She would drag you down—she would bore you. But in retrospect, a lot of us feel guilty about not being there when she needed us."

As if reaching out to say goodbye to as many friends as she could, Diane phoned people endlessly. She also became desperate to sell her work. Two weeks before she died, she made a surprise visit to gallery owner Marge Neikrug, who had just opened a shop on East Sixty-eighth Street. The two women had spoken a number of times, but had not met in person. They talked for more than five hours; Diane appeared tormented, and spoke quickly. Then, still in mid-sentence, she got up, reached for her portfolio, and left the gallery.

The following day she had dinner with her brother. They laughed and bonded as siblings do, reminiscing and sharing inside jokes. When he dropped her off at home, he could see she was utterly lonely.

That Sunday she visited two friends, Nancy Grossman and Anita Siegel, appearing unraveled. She was a forty-eight-year-old woman desperate for love and mothering. She cried for much of the after-noon, told the women she loved them, made a painting, ate some dinner, and talked about how much she liked getting her period. Then she said she wished to sleep with both of them—not in a sexual way, but more to be comforted. Cared for and nurtured. She went home, and the two never saw her again.

The morning before she killed herself, she slipped a print of Kandinsky's death mask underneath a friend's door. She had lunch at the Russian Tea Room with another. She accepted some advertising work for the *Times* and, when she ran into the photo editor, talked with him for a few moments. She ran into another friend on the street and mentioned she was thinking of moving out of New York.

On the twenty-seventh her phone rang endlessly, but was never answered. A friend was calling to make sure she would be at a speaking engagement that she had agreed to. Marvin Israel phoned a number of times and also got no response. The following day he went to her apartment and found her body in the bathroom, already decomposing.

On her desk lay her open journal. Written on the twenty-sixth were the words "The last supper." No other message was found. Israel called three friends to wait with him for the police to arrive as her neighbors argued over who would get her apartment. There was a rumor she'd taken one last picture, a self-portrait, but neither the police nor the coroner found a camera or tripod. Her uncle identified her body at the morgue. Her mother, who was staying in Florida, was phoned, and she in turn called Diane's siblings. Photographer and friend Richard Avedon flew to Paris to break the news to Doon, who was twenty-six. Her sister, Amy, was only seventeen.

Considered one of the most powerful American artists of the twentieth century, Diane was a visual reporter revealing the sad and lonely lives of people who lived on the margins of society. She altered how we looked at photographs and subjects, and what became acceptable. With an unrelenting eye for directness and emotional attachment, she was fascinated by the lost, the needy, and those searching for connection. A New York aristocrat—a word she often used to describe her subjects—she was shy, intense, and misunderstood. Her hands were often cold, even in the summer, and her eyes reflected a deadness. When she laughed or smiled, she'd hold her hand up to her mouth. Many viewed her as odd. Perhaps in the end, Diane was the loveliest of the freaks.

UNEARTHED: Wrist cutting by itself is not often effective, and few people die from the act. The reason many choose a warm bath is simple—aside from the tub's being able to contain the mess, the warm or hot *the village* VOICE, *August 5, 1971*

Diane Arbus: the mirror is broken

"A photograph is a secret about a secret. The more it tells you the less you know." Diane Arbus said that last April Fool's Day. The same statement holds true for the act of suicide. Diane Arbus took her own life last week.

It has always seemed ghoulish to me to speculate as to the motives behind a suicide, particularly so in attempts to correlate the act with an output of major creative work. Suicide can be either a confession of defeat or a hymn of triumph, an admission of one's inability to confront the pitilessness of life or a defiant refusal to tolerate it, a self-erasure or a self-definition, but it is always a private act, a final *Mind-your-own-damn-business* hurled into the teeth of the universe, and anyone who takes that irrevocable step has earned his or her privacy.

Diane Arbus slashed her wrist and bled to death in her Westbeth apartment—sometime late Monday or early Tuesday, since her diary contained an entry dated Monday, July 26. Hers was the third suicide at Westbeth, the second by a photographer. Her body was discovered by her close friend Marvin Israel, on Wednesday, July 28. Funeral services were held at Campbell's on Madison Avenue. She was 48 years old.

Diane Arbus studied with Lisette Model and earned her living as a commercial photographer, but her concern as an artist—I should say concerns, as twinned as the children in one of her most famous images—were the freakishness of normalcy and the normalcy of freakishness. She called freaks "the quiet minorities," and defined her special field of interest in photography as "a sort of contemporary anthropology," much reminiscent of August Sander, with whose work her own had considerable affinity.

In a 1967 interview for Newsweek, she said about freaks, "There's a quality of legend about them. They've passed their test in life. Most people go through life dreading they'll have a traumatic experience. Freaks were born with their trauma They've already passed it. They're aristocrats."

And, about herself, in a more recent statement: "Once I dreamed I was on a gorgeous ocean liner, all pale, gilded, cupid-encrusted, rococo as a wedding cake. There was smoke in the air, people were drinking and gambling. I knew the ship was on fire and we were sinking, slowly. They knew it too but they were very gay, dancing and singing and kissing, a little delirious. There was no hope. I was terribly elated. I could photograph anything I wanted to.

"Nothing is ever the same as they said it was. It's what I've never seen before that I recognize. . .

"Nothing is ever alike. The best thing is the difference. I get to keep what nobody needs."

Of her death, Richard Avedon, who knew her well, said, "Nothing about her life, her photographs, or her death was accidental or ordinary. They were mysterious and decisive and unimaginable, except to her. Which is the way it is with genius."

Diane Arbus's photography was Stendhal's "mirror held up alongside a highway. " That mirror is broken, by her own hand, but the mirror-with-a-memory which is the camera has allowed us to retain a few of the pieces. Perhaps someday we may even come to understand them. And thus ourselves, and her.

—A. D. Coleman

Diane's obituary in the *Village Voice*

water increases circulation, enlarges the veins, and allows blood flow to increase while slowing down clotting.

HOW IT WORKS: Cutting across one's wrist horizontally is not effective, since it's harder for blood to escape that way. Ideally, slits are made vertically, going toward the wrist from the lower part of the arm, where one is sure to cut the main artery. Generally, the first cut needs to be made with the nondominant hand, as it will be harder to slice with a weaker, already injured one. Most challenging is severing the tendons and muscle to get to the artery. Often you will see about ten to twelve hesitation marks on a victim's wrists before they can find the force and mental state to go deep enough to cause damage.

CAREER HIGHLIGHTS: Diane is known for her snapshot technique, paired with her heroic portraiture documentary style. Her early photos have a grainy quality in the style of Henri Cartier-Bresson or Robert Frank. As a newcomer in the late 1950s, she used thirty-five-millimeter cameras. During the 1960s she upgraded to the Rolleiflex medium-format twin-lens reflex, also known as a square image, which delivered a higher resolution. A waist-level viewfinder let her connect with her subjects from angles and positions that regular eye-level cameras did not. It was not unusual for her to use a flash during the daytime, which allowed her to separate and highlight subjects from the background. By 1965 she was printing her negatives uncropped and with the black border exposed.

After her death a number of retrospectives were held; the MoMA exhibit in 1972 helped cement her cultlike following. Doon was in charge of her mother's estate, and she kept a controlled eye on what work was released. In 1984, Patricia Bosworth penned *Diane Arbus: A Biography*, which depicted Diane as a tortured bohemian soul.

Mark, 1965

Mark Rothko

BORN: September 25, 1903, Dvinsk, Russian Empire (now Latvia)
DIED: February 25, 1970, New York City
AGE: 66
METHOD: Slitting his arms and OD'ing on barbiturates
DISCOVERED BY: His assistant
FUNERAL: That day, the Frank E. Campbell Funeral Home in New York housed the who's who of Manhattan's art world. Willem de Kooning, Jackson Pollock's wife, Lee Krasner, James Brooks, Robert Motherwell, Adolph Gottlieb, Philip Guston, and Jack Tworkov were among the cognoscenti sitting among dealers, collectors, friends, and family. The open casket made for an unusual and intimate experience. Mark's two children placed recordings of their father's favorite music in the coffin. A flower rested on his chest; his glasses rested on his nose. Poet Stanley Kunitz gave a eulogy. Artist Herbert Ferber gave another. The ceremony ended with Rothko's brothers reciting the Kaddish, the Jewish prayer for the dead.
FINAL RESTING PLACE: Mark had been buried for thirty-eight years in Long Island's East Marion Cemetery when his children petitioned to have his body exhumed and moved to the Jewish cemetery Sharon Gardens, in Kensico Cemetery, Valhalla, New York, where he is considered the "only notable person." Many were against the relocation, but in March 2007 a judge approved the move, and he was finally reunited with his second wife, from whom he was separated at the time of his death, and who had been relocated from a cemetery in Cleveland. His headstone merely reads, "Mark Rothko 1903–1970."

If I choose to commit suicide everyone will be sure of it.
There will be no doubts about that.
TO FRIEND AND ASSISTANT JONATHAN AHEARN

Acclaimed abstract artist Mark Rothko was found by his assistant at 9:00 AM lying on the floor in his kitchen studio, in full Jesus position, surround by a pool of blood, and with blood oozing from the crooks of his elbows. There were no empty bottles of booze or spilled pills, as one would have expected to find. No ashtray toppling over with cigarette butts or mounds of ashes. No note. No cacophony of instrumental music emanating through the East Sixty-ninth Street apartment as it normally did. Although the bed wasn't made, the dwelling was relatively clean, his pants folded neatly over a chair. He was dressed in an undershirt and long johns with blue undershorts over them. His glasses, without which he was blind, were missing or at least not part of the scene that would soon become the "Rothko suicide investigation." A double-edged razor blade with a piece of tissue wrapped over the top part was placed on a nearby shelf. Several feet away, Mark lay dead from the sizable self-made gashes in his arms and the massive number of barbiturates he'd swallowed.

The blood that had poured out of him, leaking onto the floor, covered an eight-by-six-foot area, as if he intended to become one of his own paintings, a large and dramatic abstract work of art.

Typically, Mark would have been up for a few hours, brushes in his hands, a large canvas smattered in bold, colorful strokes on the floor. But the studio was uncharacteristically quiet when his assistant, Oliver Steindecker, entered, calling for his boss. When he spotted him on the kitchen floor, he panicked and ran next door, to the apartment of painter Arthur Lidov. The two men shared a wall, and though they didn't always see eye to eye, they were still close friends.

Lidov entered, saw the oddly jaundiced body, and, rather than phone for an ambulance, as Steindecker suggested, he called the police.

The police arrived with a journalist, while an intern from Lenox Hill Hospital appeared and pronounced him dead. Lidov brewed pots of coffee in his own studio. Theodoros Stamos, Mark's accountant and an executor of his will, and Steindecker made phone calls. Rita Reinhardt, Mark's lover, and Donald McKinney arrived; they'd been scheduled to pick him up and take him to the warehouse to choose paintings for a gallery showing. Steindecker took a cab to pick up Mell, Mark's wife, and bring her back to the studio. Though they had two children and had been married for twenty-five years, the couple had decided to separate. By 12:30 PM, detectives were asking questions and a medical investigator was examining the body.

The group pieced together the information as best they could. The day before, he had had a physical and, surprisingly, was given a clean bill of health. He ran into a friend, with whom he chatted briefly about the irony that each had seen their internists that afternoon. He then had dinner with Rita at his local deli. He was nervous and angry about his meeting the following morning. At home he downed his evening meds with a glass of scotch, and Rita went home. He spoke to his brother at 9:00 PM. His neighbor, Lidov, said he'd heard nothing from Mark's studio all evening, an odd thing, since noise—the flushing of toilets, the stomping of feet, and loud music—could usually be heard through the paper-thin walls and had created an ongoing battle between the two men. Rita said she had spoken with him at midnight. What happened over the next six to eight hours no one could account for.

Mark had a pudgy face and a partially balding head. Bearing a large, bulbous nose and puffy lips, he wore thick wire-rimmed glasses and sometimes sported a mustache. He was rarely seen without a cigarette or pipe dangling from his lips or held in his hands. In photos he looked serious. Occasionally he was caught smiling.

He was often angry and depressed, frequently characterized as a melancholy artist. He carried with him abandonment issues; his father, a pharmacist and intellectual, had abruptly left the family when he was seven. Mark had two children and two failed marriages to his name. Misunderstood and underappreciated, he was constantly frustrated with the art world. Then there were his health issues. Chain smoking and heavy drinking had contributed to his high blood pressure, hypertension, and emphysema. A passion for rich food had given him gout. His eyes were failing him. He had had an aneurysm the year before, which had added to his depression and inability to work. He suffered from insomnia and anxiety. Between his internist and psychologist, he was on a messy combination of Valium, Sinequan, and chloral hydrate, a type of barbiturate.

The year 1969 had been an important one. Usually against being publicly exhibited, he agreed to a one-night-only solo show at a gallery. He recovered from a recent aneurysm of the aorta and had left his wife of twenty-five years on New Year's Day. He signed a contract making Marlborough Gallery his exclusive agent for the next eight years. He created a foundation and received an honorary doctorate of fine arts from Yale, the university he had attended for only two years before dropping out and going to New York to live as an artist. And he was commissioned by the Four Seasons restaurant to do a series of nine paintings. (He secretly wanted to create something to ruin the appetite of everyone who gobbled up the overpriced, overly rich food.) He had a number of upcoming confirmed appointments and commitments: a book writing project, a slew of important meetings.

Still, it was not surprising to friends and family when news spread that the great and powerful abstract artist had killed himself. What was startling and unrealistic was the method he chose. A known fear of blood and a queasiness in the face of violence made his method uncharacteristic. Also, for a man fond of having the last word and who possessed a passion for letter writing, his leaving no note was odd.

When police, doctors, and medical examiners inspected the scene, they were clueless as to who the dead man was. All they saw was a white man in his sixties who fell into the highest suicide category—divorced, past sixty, alone, and ill. The first autopsy report, performed by four doctors, named the cause of death as resulting from self-inflicted wounds paired with acute barbiturate poisoning. They assumed Mark had taken the pills to calm himself and numb the pain of the razor, an efficient and not unusual plan or method. The double use of pills and cutting meant he was serious about killing himself. The choice to slice his arms, though not as common as the wrists or the neck, is not as rare as some people think.

The funeral was held. The case was closed. Mark, though brilliant, would be just another devastating loss for the art world to endure.

As in a Stanley Kubrick film, a litany of odd happenings and unresolved questions surrounded his death. Friends and family never let go of the idea that things didn't add up, and in 1973, three years later, they started an investigation. The blood reports came back: Mark's blood was alcohol-free. If he had drunk Scotch at midnight, how was this finding possible? How come Dr. Mead, his cardiologist, was the only one to discover two empty bottles of pills underneath his body when a medical examiner, police officer, and intern from Lenox Hill Hospital had already reviewed the body? The reporter who trailed the police gave an incorrect version of what transpired. When the main players were again questioned about the events, each changed his story, giving conflicting reports about what he saw. Mark was a slob when it came to housework and his appearance, yet his studio seemed a little too tidy, as if someone had cleaned up. How could the myopic man have seen anything without his glasses? Why would he have wrapped a tissue around the top of the razor blade? The notice of death claimed that Mark's wife was present, when instead she had not yet been told her husband was dead.

The notes from the medical examiner's report went untranscribed until a request was made to see them in 1973. Then there was the autopsy report. Though barbiturate poisoning was listed on the death certificate, upon closer examination, it seemed that sample tissues from his brain and liver, and blood and urine, sent to the toxicology lab two days after his death, revealed no traces of barbiturates. No "basic drugs or acidic drugs" were found in his stomach—a contradiction to the autopsy report. Given these findings, a new medical examiner changed the report, citing that slashing was the only method of suicide.

The original doctor who performed the autopsy was livid when he learned that a new one had been done, and with different findings. She insisted that the lab had made a mistake. They had spelled Mark's name incorrectly, so other mistakes were possible. She said she was certain about the drug poisoning. To prove her findings were correct, she suggested that two micro-toxicological analyses be performed from saved samples kept for instances such as these. In 1975, a request was sent to Chief Medical Examiner Dominick J. DiMaio to perform such a test, but no reply was ever received. Even stranger, Mark's file disappeared from the medical examiner's office, only to reappear months later. DiMaio also refused to let a new toxicologist review the notes. There was talk that the original barbiturate tests performed were not the correct ones. And so it was decided that he had overdosed on Sinequan, bringing another change to the report.

If someone did kill him, they certainly didn't need to. The original autopsy report showed that he had only another year to live. He had senile emphysema and advanced heart disease. The doctor he saw the day before his death had misinformed the artist when he told him he was fine.

UNEARTHED: One out of every one hundred suicides employs a dual method. "People who use two methods are very serious about killing themselves and are afraid one method won't work, so they want a backup plan," says Lake County coroner Dr. Keller. "People will also use what's readily available and what they're most comfortable or familiar with." In Mark's case, that was pills and a blade. "A healthy body only needs to lose two to three liters of blood in order to die, a little more than a large soda bottle," Dr. Keller adds. "Because the body can't generate blood fast enough, and because so much is being lost from the slashes, the brain is deprived of oxygen and the person slips into unconsciousness and then death."

The least effective spot to cut is your throat, because the carotid artery is protected by the windpipe.

DID YOU KNOW: The autopsy report, number 1867, which incorrectly listed the victim's surname as "Rothknow," indicated that one hesitation mark was made, on his left arm, which bore a two-and-a-half-inch-long and half-inch-deep gash. Mark was right-handed but painted with his left, which means he used his dominant hand to make an incision on his left arm, before switching hands to cut his right arm, creating a two-inch-long, one-inch-deep slit.

CAREER HIGHLIGHTS: Completely self-taught, Mark stumbled into the art world in 1925. After spending two years at Yale, he dropped out to "wander around, bum about and starve a bit" in New York. When he went to an art class to meet a friend who was sketching nude models, he decided that this was the life for him. Known for working with six- and seven-inch brushes and massive unframed canvases covered in oil-based paint, he created simple compositions comprised of soft-edged bands of bold colors and floating rectangles. He honed this technique, one that he created, by experimenting with realism and surrealism—today people call it "sublime abstract expression." A religious man who embraced his Jewish background, he brought religious elements into his work, often asking the viewer to have an emotional and spiritual experience.

As with most artists, his work reflected his moods, and as he became more depressive, so did his art. His work grew darker, his palette narrower. In his studio, his last large canvas was an unfinished painting in reds, a color Mark said represented tragedy.

Artists Unearthed

FOR THE MASSES: For many artists, death proves to be just the start of their lifelong careers. Mark hated group shows and would turn down prestigious offers from museums and galleries. Aside from a major retrospective in 1961 at the Museum of Modern Art, he was never publicly exhibited until after his death.

One reason for afterlife popularity is posthumous resurrection. Some renowned visionaries are the subject of monumental retrospectives, and receive tributes and other honors, mostly after they've passed on. Their work tours nationally and is shown in every major museum worldwide, bringing their names and their work to the masses, many who didn't have the opportunity to view them before.

Starting in 1959, Diane kept an assortment of writings: small black spiral notebooks filled with witty notes and scribbles, postcards, to-do lists, quotes from books she was reading, funny things people said, thoughts and musings, and the personal life stories her subjects told her. Thirty-plus years after her death, these notebooks were included in *Revelations*, a show organized by the San Francisco Museum of Modern Art in 2003. The exhibition traveled throughout Europe and the United States. More than seven million viewers came to see it in North America.

In many cases, the families who become the executors of dead artists' estates are the ones responsible for reassembling and organizing their loved ones' work. Van Gogh's sister-in-law, Johanna, spearheaded a mammoth project that established his reputation as a brilliant and overlooked artist. After van Gogh's brother Theo died, Johanna inherited the entire collection. Her husband had deemed the work worthless and suggested burning his paintings, letters, and drawings. Relatives who were offered some of his works echoed similar sentiments, and turned their noses up at them. Today those pieces are worth millions.

After moving from Paris to Amsterdam, Johanna led a tireless campaign, organizing exhibitions, cataloguing his work, and, for the first time, translating his letters from French into Dutch. It took a decade, but her brother-in-law's name soon became equated with artistic genius. In 1905 a massive showing of his work took place at the Stedelijk Museum, in Amsterdam. After Johanna's death, her son took over the family calling: he established a van Gogh foundation and had the Dutch government create a museum dedicated to his uncle. In June 1973, Queen Juliana of the Netherlands opened the Vincent van Gogh National Museum. With this event setting in place a domino effect that made his work available to a wider audience, more and more people became aware of his greatness. Finally, he received the artistic acclaim and accolades he deserved. Memorial exhibitions were also mounted in Brussels, Paris, The Hague, and Antwerp. In the early twentieth century, the exhibitions were followed by vast retrospectives in Paris, Amsterdam, Cologne, New York City, and Berlin. These had a noticeable impact on artists and amateurs alike. Today, van Gogh's work has established record prices broken by few.

THE RUMOR MILL: If the way these visionaries died didn't cause enough of a shout throughout the art world, then the rumors over their lives did. Though their work was magnificent, the stories created in their absence propelled their popularity and glamour.

Diane supposedly set up and snapped a self-portrait in the last moments of her life; it has never been found. Many feel that her friend Marvin Israel, who reached her apartment first and found her body, removed the evidence of it. Because of her photographs, Diane had already created controversy; the rumor of the missing portrait created even more secrecy and intrigue.

When Mark died, 798 paintings, valued at $32 million, were secretly signed and severely undersold to the Marlborough Gallery by the three executors of his will—Theodoros Stamos, Bernard Ries, and Professor Levine, each of whom collected exorbitant commissions and divided the proceeds. In 1975, in a case called the "Watergate of the Art World," Mark's children, Kate and Christopher, nineteen and twelve, respectively, sued the trio for falsifying evidence and funds, as well as for negligence and conflict of interest. The men were found guilty, removed as executors, and fined, along with the Marlborough Gallery, $9.2 million.

In 1999, rumors surfaced that the seventy paintings van Gogh created in the last ten weeks of his life—such as *The Church at Auvers* and *The Cornfield*—were not all done by him, but rather by his doctor, Paul Gachet, whom van Gogh had been seeing and with whom he'd quarreled the day he shot himself. This would certainly put into perspective why he brought a gun to the doctor's home and flew into a rage (supposedly over his painting not having been hung). Legal action was taken, and curators of an exhibition hung van Gogh's and Gachet's known work side by side in the hope of telling them apart. The paintings were also subjected to twelve months of rigorous chemical and X-ray analysis.

DID YOU KNOW: Cartoons and comic strips have also highlighted the real threat of suicide, though they've added a humorous element.

Suicide Sheik, released in 1929, told of Lucky Rabbit Oswald, who is snubbed by his ladylove when he proposes to her. Heartbroken, he sees suicide as the only answer. Of course, he doesn't go through with it, and when his ex-girlfriend's home catches fire, he rescues her and the pair is reunited. Walt Disney took the idea of suicide one step further, illustrating a heartbroken Mickey Mouse trying an array of methods, none of which proves successful, in a one-week segment of *Mr. Slicker and the Egg Robbers*.

"Sometimes comic strips that seem innocent in one era can seem startling, or at least controversial, in another," says David Gerstein, archival editor with Gemstone's Walt Disney Comics. "There was a vogue, in the 1920s and early Depression era, for gags based on the concept that backfiring suicide was funny. Then as now, of course, suicide was a grim fact of life; in 1930, however, it was far less common among youth. The [Disney] story's treatment of suicide is proactive: its climax, in which Mickey firmly decides to stay alive, can easily be read as an authorial attempt to smarten up younger readers."

Seven
Powerful People

Forward (and in one case, backward) thinkers—inventors, math-ematicians, philosophers, holders of political seats, and those who opposed them—are grouped in this section, mostly for the contribution they made to world history. Whether creating the first computer, presiding over a country, discovering the cause for repressed sexual desire, or protesting a war, the people in this group—with their passion and devotion, their unique findings, political stances, or commitments—advanced or destroyed society and forever altered the world.

Beginning with the ancient sovereigns and continuing into the present, suicide has always been a favored option, not to mention an easy escape from punishment. It's an ongoing debate—was the classical Greek philosopher Socrates, who is credited as one of the greatest thinkers and a founder of Western philosophy, forced to drink hemlock as a way to end his life—or did he choose to do it on his

Socrates holding a cup of hemlock

own? There are many arguments pointing to both sides of the suicidal coin. His historic decision came at a time when suicide was still considered a noble act. If he chose to end his life, he'd be the first documented case not tied to the Bible.

Details about Socrates's life are sketchy. Our knowledge of the man born in 469 BC, and who started out as a sculptor, is based on writings by his students and contemporaries. The main sources include the *Dialogues* of Plato, the writing of Xenophon, and Aristophanes's plays. At seventy years of age, Socrates clashed with Athenian politics and society. Charged with "refusing to recognize the gods recognized by the state" and of "corrupting the youth," he was tried in front of a five-hundred-person jury made up of his peers. The majority ruled in favor of death, which under Athenian law meant drinking hemlock. Other scholars believe that Socrates was asked to select his method of execution, and that hemlock was his death drink of choice. Some sources say that once the liquid was drunk, Socrates was to walk around until the poison did its job and a gradual paralysis of the central nervous system occurred. Others say he spent his final hours in a jail cell.

Unlike creative people—actors, writers, musicians, and artists—powerful people, or people in power, seem to choose suicide as an escape from punishment or an inability to admit or accept failure. Socrates was one of many who "cheated the hangman," the ultimate act of defiance, an act that deprived emperors, kings, and presidents of the ability to execute their "victims."

There isn't a continent or high-powered country that hasn't been affected by the self-killing of a ruler. When politicians end their lives abruptly, it changes the face of the country they govern, and in some cases, the political scope of the world. Leaders in office often seek a quick out, killing themselves for a variety of reasons, of which illegal actions are the most popular.

Ricardo Bordallo was elected governor of Guam twice. In 1987 he was convicted on corruption charges, including bribery, witness tampering, and obstruction of justice. Nicknamed Section 8—a military term for "insane"—for his repeated attempts to get out of military service, Bordallo committed suicide in 1990. On the morning he was to report to prison, he wrapped himself in the Guamanian flag, chained himself to a statue, and put a bullet through his head, all in the middle of rush-hour traffic.

Shortly after his inauguration for a fifth term, New York City politician Donald R. Manes, the borough president of Queens, committed suicide in 1986. He made two attempts. The evening of the first, he attended a dinner party, was seen leaving in his car, and was found the following morning dripping with blood from slashes to his wrists. Originally, he said two men had attacked him, but he later recanted the statement, admitting that the injury was self-inflicted. Weeks later, news surfaced that he was involved in large kickback schemes and other means of corruption. While on the phone on hold with his shrink, who had been momentarily called away, Manes decided to extract a kitchen knife from a drawer and plunge it into his chest. His daughter found him on the floor in a pool of blood.

In 1891, after being accused of embezzling $20,000 from an estate of which he was the executor, New York lawyer Charles A. Binder fled his hometown, drifted around New Jersey, bought a gun, and finally checked himself into the Sheridan House Hotel under the alias "John Roth," where he shot himself in the temple. In the note he sent his wife, he complained of nervous strain, that he feared arrest and had grown weary.

In 1954, Lester Callaway Hunt, who was both the governor of Wyoming and later a U.S. senator, killed himself while still in office after being blackmailed. His son had been arrested for soliciting a male prostitute, and another senator had demanded that Hunt step down so a Republican could take his seat. If he refused, voters would find out about his son's secret. Eleven days after announcing that he would not run again, Hunt shot himself in his Senate office.

Pressures of the job caused Massachusetts U.S. representative George Pelton Lawrence to jump from an eighth-floor window at the Belmont Hotel in 1917, and U.S. representative Marion Anthony Zioncheck to jump from the window of his campaign office in Seattle in 1936. Ironically, his body struck the pavement directly in front of the car his wife was sitting in.

And family woes: By 1965, Balfour Bowen Thorn Lord, a Democratic politician from New Jersey, had two failed marriages. Depressed by his recent separation from his second wife, he strangled himself with an electrical cord at a friend's home.

Physicians and inventors are not immune, either. British forensic pathologist Sir Bernard Spilsbury, considered the foremost forensic "detective-pathologist" of his era, was noted for setting the standard for forensic investigations and speaking at the most sensational trials of the early twentieth century. Suffering from depression as a result of a failed marriage, crippling arthritis in his hands, and the death of his sister and two sons, Spilsbury resolved to kill himself. On a cold night in December 1947, he dined alone and then went to his laboratory, where he gassed himself to death with a Bunsen burner.

Edwin Armstrong, an electrical engineer and inventor, gained notoriety for creating FM radio, among other things. In 1954, financially drained and depressed over an FM patent dispute, he dressed in his coat and hat and dove out of the window of his thirteenth-floor Manhattan apartment.

Austrian physicist Ludwig Boltzmann was famous for discovering statistical mechanics—which describes how the properties of atoms determine the properties of the substances they form. Having attempted suicide twice before, and having a history of depression, the doctor was particularly resentful over the scientific establishment's rejection of his theories. During a vacation in the coastal town of Duino, in Italy, he hanged himself in his hotel while his wife and daughter were out swimming.

Feeling his career had peaked despite his huge success with his invention of nylon, and holding more than fifty patents, American chemist Wallace Hume Carothers fell into a melancholy state that was exacerbated by the death of his sister. And though his wife was two months' pregnant with their first child, in April of 1937 he checked into a hotel in Philadelphia and made himself a cocktail: lemon juice laced with potassium cyanide.

Unlike Adolf Hitler, who chose two methods, Hitler's wife, Eva, Sigmund Freud, Alan Turing, and Abbie Hoffman all shared one thing: their suicide method. As different as their backgrounds, interests, and even the reason for killing themselves, no one drifted off to sea with the intention of drowning or impulsively leaped off a bridge or gravitated toward the sharp edge of a razor. Whether overdosing on sedatives, lacing an edible object with cyanide, or having a doctor administer too much morphine, all chose poison as their method.

To many, poison seems like a simple, non-messy, self-contained, almost pleasant way to go. When administered correctly, poison can drug you into a heavy sleep.

"The point is slowness. Poison lets you glide your way into a painless, easeful death, as Shakespeare called it," says cultural historian Leo Braudy. "Then perhaps there's always the element of hesitancy, since poison can often be reversed. Hemlock is slow, the bite of an asp is slow, morphine is slow. Arsenic is generally much more immediately painful."

Adolf Hitler and German officials inspecting damage from dropped bombs in a German city, 1944

Adolf Hitler

BORN: April 20, 1889, Braunau am Inn, Austria-Hungary
DIED: April 30, 1945, Berlin, Germany
AGE: 56
METHOD: Gunshot and cyanide poisoning
DISCOVERED BY: Otto Gunsche, a unit leader of the Nazi Party, in Hitler's bunker
FUNERAL: As ordered, Hitler and Eva's bodies were carried up to

Berlin's Chancellery gardens, doused with gasoline, and burned.

FINAL RESTING PLACE: There has been massive controversy over what happened to their remains, and their whereabouts are still a mystery. Most believe the couple were swept into a canvas tarp, placed into a shell crater, and buried.

> *I have not come into this world to make men better,*
> *but to make use of their weaknesses.*
>
> QUOTED IN THE ESSAY "THE MIND OF HITLER," BY H. R. TREVOR-ROPER

The small civil ceremony takes place just before midnight in a private suite. Only a few comrades and staffers witness the couple saying their vows in front of a city official who's been drafted from the Volkssturm military unit. Champagne, a delicacy during wartime, is brought out; the popping of the cork is extra loud as it reverberates off the fourteen-foot-thick concrete walls of the underground bunker that Hitler has called home for the past several months. In the bride's excitement, she accidentally writes the wrong name on the marriage certificate, crosses out the *B* for *Braun*, and enters "Eva Hitler née Braun" instead. After glasses are drained and congratulations bestowed, the cheerful banter turns to talk of suicide. Hitler and Eva agree to die in twenty-four hours.

He never intended to wed her, but given the turn of events, he feels that her devotion and friendship should be rewarded.

"Although during the years of struggle I considered myself unable to take on the responsibilities of marriage," Hitler said, according to Guido Knopp's book, *Hitler's Women*, "I have now decided, shortly before the end of my life, to wed the woman who, after many years of true friendship, freely came to the beleaguered city of Berlin, in order to share my destiny. As my wife, at her own request, she will join me in death."

The couple meets in 1929 at Hitler's photographer's studio in Schellingstrasse. He is forty, she, seventeen. During their courtship, their dates are arranged in secret, since her father dislikes him. Always in control, Hitler decides when and where they will meet, and what they will do.

A man of his word, Hitler, along with his new wife, commits suicide the following day. After a gunshot is heard, Otto Gunsche, a unit leader of the Nazi Party, rushes into Hitler's private room and finds the newlyweds dead. Hitler's bloody body is crumpled up, his head hanging toward the floor. Blood gushes from his temple, runs down his clothing, and pools on the carpet near a fallen pistol. Eva sits on the other side of a small couch, her legs curled under her. Thanks to the cyanide pill she took, she, too, is dead. Though a gun is by her side, it remains unfired.

Adolf Hitler, one of the most hated men in history, ruled Germany for twelve years and was responsible for the deaths of millions in World War II—most specifically the Jews, a group for which he harbored deep contempt.

His signature inch-wide black mustache, swastika armband, crisp uniform, and "Heil Hitler" salute became synonymous with fear, prejudice, and war.

Fanatical about Germany and his German roots, he decided to join the Bavarian army after Germany entered World War I. He served in France and Belgium as a runner, and was awarded the Iron Cross twice as a result of his actions. He was also given the Wound Badge after being shot in his left leg during the Battle of the Somme. He joined the German Workers' Party in 1919, and by 1921 he'd become the leader of the group, renaming it the National Socialist German Workers' Party (better known as the Nazi Party). During this time he immersed himself in his three passions: art, Germany, and a hatred for the Jewish people. Two years later he led an unsuccessful attempt to overthrow the ruling German Weimar Republic. He was sent to prison, but rather than wallow in his failure, he wrote his manifesto *Mein Kampf,* or *My Struggle.* Less than a year later he was released, and quickly became a populist spokesman for poor and nationalistic Germans. His explosive power grew, and in 1993, after single-handedly suspending the constitution, suppressing all political opposition, and putting the Nazis in a position to reign over Germany, he was named chancellor. His next act was to create both the Gestapo, a secret police force, and the "Protection Squad," also known as the Schutzstaffel, or SS. This was followed by his devising the "Final Solution"—the systematic murder of Jews in occupied territories and the creation of the infamous concentration camps, where millions of Jews were enslaved and eventually killed through a variety of torturous and monstrous methods.

By August of 1934, Hitler had become Führer of Germany, making him the most unstoppable force in history—until England declared war on his government.

By 1942, the Final Solution was fully instituted, but the tables turned when Germany lost the pivotal Battle of Stalingrad. Though a plot to assassinate Hitler failed in mid-1944, it scared the Führer enough to cause him to retreat to his bunker beneath Berlin, fearing for his life as the Allies marched into Germany.

On January 16, 1945, Hitler and Eva, three-dozen medical and administrative staffers, the propaganda minister and his wife and their six children, and others close to Hitler, moved into the thirty small rooms under Berlin's Chancellery gardens.

During his four-month stay in the bunker, Hitler, a severe claustrophobic known for his need to constantly command his environment and move about freely, became a paranoid madman, spitting into fits of rage, throwing tantrums, and living like a rat trapped in a high-end life-size maze.

The following timeline tracks the last two weeks leading up to his suicide:

April 19: Soviet armies force through German lines and encircle Berlin.

April 21: Hitler's birthday. In celebration, he takes his last trip outside, where he walks in the gardens. He also orders German army formations to counterattack Soviet troops, who are closing in.

April 22: During a three-hour military conference, he becomes irrational and erratic, spewing that his army, aides, and friends are corrupt liars, and are the reason for his failures. Disgusted with their lack of commitment, he offers those who wish to do so, a chance to leave. Many accept.

April 27: The Soviets bomb the Chancellery buildings, and Hitler sends a frenzied telegram to Field Marshal Keitel insisting that more troops be sent to him.

April 28: British news reports says his close friend and Nazi Party leader Heinrich Himmler has sought negotiations with the Allies, and has offered to surrender German armies in the west to General Eisenhower. Filled with rage, an uncontrollable Hitler orders that Himmler be arrested and, out of revenge, that Himmler's personal representative (Eva's brother-in-law) be shot.

At 4:00 AM, he dictates his last will and a two-part political testament reiterating sentiments expressed in *Mein Kampf*, blaming the Jews for everything, including World War II. Feeling sentimental, and with Eva being one of his few remaining minions, he decides to marry her. The couple has a wedding breakfast in the wee hours of the morning.

April 29: Hitler is informed that Mussolini has been captured and shot, and his body hung upside down and then thrown into the gutter. Fearful that his prized cyanide capsules, which he's been hording, have been tampered with, he feeds one to his favorite dog to see if it works. The dog dies, and Hitler distributes the poison to his secretaries to use if the Soviets, who are a mere mile away, should break through to the bunker. He then apologizes that "he did not have better 'parting gifts' to give them." The staffers break out in dance, cheering and singing as if at a holiday party.

Over lunch, Hitler says the safest way to kill yourself is to put the barrel of a gun into your mouth and pull the trigger. "The skull is shattered in pieces and death is instantaneous." Eva declares her desire to be a beautiful corpse, and pulls from the pocket of her dress a cyanide pill. She spends the rest of the day writing farewell letters.

April 30, about 2:30 AM: Hitler gives a goodbye speech to his remaining staffers. Eva emerges wearing her husband's favorite low-cut black dress. Her hair is perfectly styled.

April 30, 2:00 PM: Informed that the Soviets are just a block away, Hitler eats a vegetarian lunch, then orders his chauffeur to deliver two hundred liters of gasoline to the Chancellery gardens. Dressed in a field-gray jacket, black trousers, and matching shoes, he joins Eva in saying their final adieus before entering their private room.

The rest is a blur: The pop of a gunshot. The waiting to be told what to do. Hitler's bloody body sprawled on the sofa, the Walther PPK 7.65-millimeter pistol by his feet. Eva's lifeless body slumped away from his. The booming sound of bombs being dropped by the Soviets.

In true Hitler style, he leaves specific instructions describing how his and Eva's bodies are to be disposed. Wrapped in blankets, they are carried out to the gardens by Otto Gunsche and two SS men. The doors leading to the gardens are locked and the bodies laid in a shallow depression in the earth. Cans of petrol are poured over the corpses and they are set on fire. A last salute is given. That evening their charred remains are swept into a canvas tarp, put in a shell hole outside the bunker exit, and covered with earth.

May 2, 1945, front page headline on Hitler's death in the *News Chronicle*

UNEARTHED: Hitler demonstrated affection for the young girls who idolized him. He took tremendous pleasure in educating eighteen- to twenty-year-olds, seeing them as "malleable as wax." Like most things he touched, these women often met unfortunate ends.

1. In 1926, he courted a pretty, blond, seventeen-year-old draper's assistant from Berchtesgaden named Maria Reiter. Though kissing was the extent of their relationship, she was infatuated with him, and when she realized they might not marry, she tried to hang herself from a doorpost. Her family stopped her just in time and she survived.

2. In 1931, Geli Raubal, Hitler's niece and supposed lover—with whom he had an obsessive, controlling, and horrendous relationship—shot herself when he ordered her to stay at his apartment and not go to Vienna while he was away. In her honor, he created a "Geli Room" in his Berghof residence, which he filled with portraits and busts of her. On several occasions he locked himself inside with a loaded pistol and threatened suicide. Reports surfaced that he had killed her or arranged to have her murdered, but these proved unfounded. No autopsy of Raubal was conducted.

3. In 1932 and 1935, Eva Braun attempted to kill herself over Hitler. On Halloween, after not hearing from him for three months, the twenty-year-old wrote a farewell letter, then shot herself in the chest with her father's pistol. Her sisters found her covered in blood. With medical treatment, she survived, but another attempt was made three years later. Having swallowed thirty-five Phanodorm sleeping tablets, she was again saved by her sister.

4. In 1939, hours after World War II started, Unity Mitford, a well-known British aristocrat who was infatuated with Hitler, sat on a park bench in Munich and put a bullet through her head. She survived the incident but died a few years later from the injuries.

THE MISSING ASHES: Hitler's death and the whereabouts of his remains are still a mystery. An official British inquiry into how he killed himself was announced on November 1, and it was found he only shot himself; later it was proved he bit into a cyanide pill before using the gun. Some believe he stuck the pistol in his mouth, because in existing photos of his corpse, his mouth and chin were twisted to one side. Others say he shot himself in the temple, because his head was leaning forward and slightly angled to the right. There was also a dark blotch located above the outer edge of his right eye.

There are countless theories regarding what happened to his remains. Most accepted is that the Russians found the burned corpses of Hitler and Eva and reburied them. Some believe the Soviet Army counterintelligence buried Hitler in various places around Germany as they made their way back east. Some believe the KGB kept the two buried at a military base in East Germany from the late 1940s to the early1970s, at which time the bodies were exhumed and completely destroyed. In 1993 the Russian government insisted they had fragments of his cranium, but noted Hitler biographer Werner Maser claimed it was a fake. Pieces of his teeth and bridges materialized as well, and were positively identified by Käthe Hausermann and Fritz Echtmann, assistants to Hitler's dentist, Dr. Hugo Blashke. Most outlandish is the idea that Hitler had "doubles" and that it was one of them who was shot, carried up to the gardens, and burned, while the real Hitler escaped to Argentina. Those close to him suggested that he headed to the Middle East, where he was still well liked. For a period after the war, random "Hitler sightings" occurred in Spain, Holland, and Scandinavia, among other places.

THE BUNKER: The Führerbunker, which translates to "shelter for the leader," was Hitler's complex of subterranean rooms, made famous first for being his headquarters, then as the place where he and Eva killed themselves. Located more than fifty feet below the Chancellery gardens and protected by approximately four meters of concrete, the two buildings comprising the Führerbunker were connected by sets of stairs with emergency exits leading to the gardens. The complex was built in two different phases: part one in 1936, and the other, nine years later. His private rooms were in the newer, lower section. By February 1945, the Führer had decorated his new home with high-quality furniture. Like any fallout shelter, the bunker was well stocked with supplies and necessities. During his last week, Hitler is said to have drunk ten to sixteen cups of tea per day.

CAREER HIGHLIGHTS: Hitler didn't start out a Jew-hating madman. It grew slowly, like an illness that lies dormant for many years, quietly brewing. Originally, his dream was to be an artist. He moved to Vienna with the goal of entering the Vienna Academy of Art, but he wasn't accepted; the instructor grading the entrance exam felt that his drawing abilities were "unsatisfactory." A passion for architecture led him to the Vienna School of Architecture, but he was rejected once again, this time due to his spotty high-school record. Refusing to give up, he lodged in a men's hostel in Vienna, turning out water colors and selling postcards he painted of the city, until his political interests and endeavors overshadowed his attempts at art. His follow-up to *Mein Kampf*, which was released in two volumes in 1925 and 1926, was *Zweites Buch*, or *Second Book*, and delved in foreign policy and his predictions for the future, but he decided against publication. His other major failure happened in 1930. Hoping to capitalize on the momentum from the increased number of parliamentary seats created for the Nazis Party, the second largest party in Germany, he ran for president, but lost. Refusing to accept failure, and obsessed with power, he ran a second time, and became the first politician to campaign by aircraft. Conceived by future Nazi propaganda maestro Joseph Goebbels, the campaign was dubbed "Hitler over Germany."

Sigmund Freud and his daughter Anna

Sigmund Freud

BORN: May 6, 1856, Moravian city of Freiberg (now in the Czech Republic)

DIED: September 23, 1939, Maresfield Gardens, London

AGE: 83

METHOD: Morphine overdose

DISCOVERED BY: His friend Dr. Max Schur

FUNERAL: Three days after his death, Freud's body was cremated at Golders Green Crematorium in London. Austrian refugees, including the author Stefan Zweig, attended the ceremony.

FINAL RESTING PLACE: Freud's ashes were placed in an ancient Greek urn that he had received as a present from Marie Bonaparte. After his wife's death in 1951, her ashes were added to the urn.

Towards the actual person who has died we adopt a special attitude:
something like admiration for someone who has accomplished
a very difficult task.
FROM HIS *THOUGHTS FOR THE TIMES ON WAR AND DEATH*, 1915

An older man dressed in a suit and tie, sporting round specs and a white mustache and a beard that runs ear to ear, sits nodding, swathed in a billowing haze of cigar smoke, while his patient lies on a couch free-associating—uttering anything that comes into his mind. Notes might be made during the session by the great and highly intellectual Father of Psychotherapy as he sits a few feet away, completely hidden from view.

Two months before the end of Sigmund Freud's life, riddled with jaw cancer and in chronic pain, he continued to see four patients. In his final days, no longer able to work, his speech and hearing badly impaired, Freud requested that his bed be brought to his London study so he could be near his books, his desk, and his antiquities. With his mouth full of holes, wearing a prosthetic jaw, and emitting a rancid odor so pungent a mosquito net had to be placed over his bed because the smell attracted flies, he knew his life was over. And that someone needed to help him end it.

The first of seven children born to Austrian Jewish parents in Vienna, Freud showed early signs of brilliance, and was thus favored by both parents. At college age, he planned to study law—one of the few areas still open to Jews—but found himself working on a research project at the University of Vienna, where he studied the testicles of eels. Frustrated by the lack of success that would have gained him fame, he changed his area of medicine, to a concentration in dynamics of psychology of the mind and its relation to the unconscious. He applied the fundamentals of chemistry and physics to the mind, forming his first basic models. After graduating and receiving his M.D., he traveled to Paris on a fellowship to study under Europe's foremost neurologist, Jean-Martin Charcot. In the mid-1880s, he married Martha Bernays and opened his own medical practice, specializing in neurology, where he experimented on his patients with hypnosis. From their uncensored spewing of thoughts and feelings, he found that his patients had a higher rate of successful breakthroughs. He called this "the talking cure," which was widely seen as the basis of psychoanalysis.

During his years in practice, he created the clinical practice of psychoanalysis, which included unprecedented breakthroughs in thinking on sexual desire, free association, the theory of transference, and the importance of dream interpretation. In essence, he redefined the human mind and behavior while paying specific attention to logic, lucidity, and lunacy.

His early work concentrated on the conscious, unconscious, and preconscious, while his later findings probed the psyche, dividing it into id, ego, and superego. He was fascinated by dreams—calling them the "royal road to the unconscious"—and repression, be it of sexual or merely painful memories, insisting that people unaware of their buried memories and traumatic experiences were unable to address them.

He started his days at 7:00 AM, saw patients from 8:00 until noon, stopping for lunch with his family. A daily jaunt around the neighborhood for an hour followed. More patients were seen until 7:00 PM. After dinner was a lively card game with his sister-in-law. Another walk was taken around the neighborhood, during which he would stop to read the paper and smoke at a café. At nine or ten, he could be found in his study working on his lectures and manuscripts or writing letters until after midnight. He kept a journal obsessively throughout his life, including exact logs of his thoughts, routines, and even cigar purchases. For whatever reason, he burned his personal papers in 1885 and again in 1907, so little is known about this part of his life, especially after he graduated.

In 1884, Freud was introduced to cocaine by a friend, who recommended the potent powder for the treatment of nasal reflex neurosis. In his paper "On Coca," he praised the drug and explained its virtues. Unaware of its addictive qualities, he suggested his friends and family members use what he called his "magical healer." His reputation took an extreme plunge when, after endorsing the drug, reports of addiction and overdosing began to spread around the world. Though he felt bad about this, he continued to use the drug while simultaneously being plagued by migraines, nightmares, heart trouble, and severe anxiety.

In his forties, Freud confronted his numerous psychosomatic disorders, acknowledging that he himself possessed exaggerated fears of dying and other phobias. While exploring his own memories, dreams, and personality issues, he unearthed the hostility he felt toward his father, who had died in 1896, and admitted having feelings for his mother; the resulting theory became known as the Oedipus complex.

He published numerous papers and books within the first few years of the 1900s, including *The Interpretation of Dreams* and *The Psychopathology of Everyday Life*, in which he theorized that forgetfulness or slips of the tongue—now nicknamed "Freudian slips"—

had important and symbolic meanings. In 1902, he was appointed professor at the University of Vienna and began to gather a devoted following. As his following grew, so did the number of his theories. Though he never looked into his own relationship with oral fixation, he got the most pleasure from smoking cigars incessantly. He consumed twenty daily, which led to his sixteen-year-long battle with jaw cancer.

Diagnosed in 1923, he spent the following decade and a half undergoing more than thirty operations to remove tumors and be fitted for several mouth prostheses, each of which he called "the monster." His addiction to cigars was so intense that he would use a clothes peg to wedge open his jaw and insert the rolled tobacco.

In 1930, he received the Goethe Prize in appreciation for his contribution to psychology and German literary culture. Years later, Nazis would burn his books when Hitler took over Vienna in 1938. During the war, his passport was confiscated, but his celebrity status allowed him special privileges. At eighty-one years old, he, along with his wife and Anna, his favorite daughter turned caretaker, fled to England, where they would be safe.

As Hitler's power blazed through parts of Europe, cancer did the same to Freud. Aside from the numerous operations, there were forty-eight office visits, one biopsy, two cauterizations of new lesions, and constant experiments to improve his prostheses, which were used to replace cancerous tissue. As the years wore on he grew sicker; and since antibiotics were not yet available and he refused pain medication, feeling it clouded his mind, little could be done to improve his situation.

Late in July 1939, two more lesions developed on his jaw, and he underwent surgery at the London Clinic. Still, more lesions appeared. The skin over the cheekbone became gangrenous, eventually creating a hole between the oral cavity and the outside. The rancid odor emanating from his mouth was so bad even his dog wouldn't go near him. Too frail for surgery, he underwent radiotherapy, which created unpleasant side effects: fatigue, dizziness, headaches. He lost his beard and bled from his mouth.

His biggest champion was friend and doctor Max Schur, who became Freud's personal physician in 1929. It was Marie Bonaparte who persuaded Freud to make Dr. Schur his personal physician. Their relationship was so close that Freud included Schur and his family in the "Freud" party when they were given safe passage to London.

In September, bedridden and unable to read or work, he took Dr. Schur's hand and asked if he recalled the promise he'd made. "My dear Schur," he said, "you certainly remember our first talk. You promised me then not to forsake me when my time comes. Now it's nothing but torture and makes no sense any more." The doctor was a man of his word, and after Freud suggested Schur confer with Anna, it was decided.

For more than seven hundred years, English common law had punished or disapproved of suicide and the assisting of it. But in the mid- to late 1930s, euthanasia societies were formed in England and the United States. On September 21, Schur injected Freud with three centigrams of morphine (the normal dose for sedation was two), and he slipped seamlessly into sleep. Repeated injections were administered when he began to stir, with a final injection the following day. This caused him to lapse into a coma, from which he never woke. He died at 3:00 AM on September 23. Dr. Schur was never charged with a crime, even when he divulged the details of his friend's death in the book *Freud: Living and Dying*.

In *Beyond the Pleasure Principle*, he concludes that all humans entertain an unacknowledged death wish, a universal desire to commit suicide. "The aim of all life is death," he writes. Freud, who was often preoccupied by death, devised the death drive theory in 1920, stating that one is born with a death instinct, which makes it impossible to imagine one's own end. Sigmund Freud proved his own theory wrong when he asked Schur to kill him. As he was clearly able to

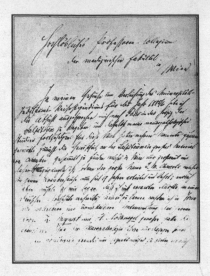

Freud's letter detailing a trip from Paris to Berlin in 1886

understand, if not at least imagine, the outcome, there was nothing unconscious or repressed about his decision to take his own life. It was just one more situation he had complete control over.

UNEARTHED: *Euthanasia*, which means "good death" in ancient Greek, translates in modern language to the act or practice of taking one's own life or helping another person to take his life, usually in cases of terminal illness. Though many still see euthanasia as murderous, others view it as merciful. The first known anti-euthanasia law in the United States was passed in New York in 1828. After the Civil War, voluntary euthanasia was promoted by advocates, with its support peaking toward the turn of the century. Michigan physician Dr. Jack Kevorkian—with the soapbox statement "dying is not a crime"—

has advocated for the rights of the terminally ill while assisting in more than 130 suicides, mostly from 1990 to 1998, the year Oregon legalized assisted suicide. Twelve months later, Dr. Kevorkian was jailed for second-degree murder, spending 1999 to 2007 in prison. His favored method is a mask attached to a tank of carbon monoxide, which allows patients to open a valve that starts the gas flowing. Death usually occurs in less than five minutes.

DID YOU KNOW: Narcan restarts a person's breathing after they've overdosed on opiates such as morphine and codeine. It is the drug paramedics use to revive people. Yet its magical healing effects last for only about an hour. Once it wears off, the person can OD again.

DID YOU KNOW: In 1993, the book *Final Exit*, a suicide guide for the terminally ill, hit the shelves. For the sick and suffering, it suggested asphyxiation as the best means for killing oneself. That year, suicides by that method in New York City rose from eight to thirty-three. Eerily, a copy of the book was found at the scenes of nine of them.

CAREER HIGHLIGHTS: All of Freud's findings, theories, and writings revolutionized and influenced art, literature, social and psychological thinking, and behavior. He discovered the existence of the unconscious mind, the "talking cure," the concept of the Oedipus complex, defense mechanisms, Freudian slips, and dream symbolism.

He published fifteen major works, including *Studies on Hysteria, Three Essays on the Theory of Sexuality, The Ego and the Id,* and *Civiliza-*

tion and Its Discontents. Eleven books of letters and correspondence were also published, including *The Freud/Jung Letters, The Complete Correspondence of Sigmund Freud and Karl Abraham,* and *The Correspondence of Sigmund Freud and Sándor Ferenczi,* Volumes 1, 2, and 3.

Freud's therapist's couch, a gift from a Viennese patient, is among his belongings located at 20 Maresfield Gardens, London, the home where he died, now a museum. Another museum is located in Vienna. The Viennese museum celebrates the famous couch through paintings and photographs of celebrity couch sprawlers.

His theories and research methods have always been controversial. His contributions to psychotherapy have been extensively criticized, and his sexual discoveries have offended. He has been cited for his dishonesty and for not acknowledging the many "mind men" who paved the way for him. He is also often called a mythomaniac, with a compulsion to embellish the truth or tell outright lies. Industry pioneers claimed that he knew his patients were not always healed in the ways he attested, and that he wrote their confessions of being cured and then had them sign them. Others were sure he lied about their treatments in order to make himself and his teachings appear legitimate. Others felt his theories contained no scientific evidence whatsoever. Or, as the English psychologist Hans Eysenck concluded, "[O]ver 80 years after the original publication of Freudian theories, there still is no sign that they can be supported by adequate experimental evidence, or by clinical studies, statistical investigations or observational methods."

Alan, 1951

Alan Turing

BORN: June 23, 1912, London, England
DIED: June 7, 1954, Winslow, Cheshire, England
AGE. 41
METHOD: Poison
DISCOVERED BY: His housekeeper

FUNERAL: There was no memorial; instead, Turing was cremated on June 12 at the Woking Crematorium. His mother, brother, and friend Lyn Newman attended the ceremony.

FINAL RESTING PLACE: His ashes were scattered near his father's in the Woking Crematorium gardens.

One day ladies will take their computers for walks in the park and tell each other, "My little computer said such a funny thing this morning!"
IN AN INTERVIEW WITH *TIME* MAGAZINE, MARCH 29, 1949

The police were at the door, and Alan Turing let them in. After all, he had called them. Earlier that morning he'd awakened to find he'd been robbed, again, most likely by the friend of the nineteen-year-old man he was sleeping with. He felt invaded, unsafe. And so, when they rang the bell, the shy, sensitive, and naïvely honest mathematician greeted the police warmly. Once they were inside his home, he told them about the small, but significant missing items: a shirt, fish knives, a pair of trousers, his shaver, a compass, and cash. Though the police were interested in the crime, another had been committed—and Alan was guilty of it, unabashedly admitting to it from the start.

The 1950s in the UK was a tumultuous time. England seemed especially dreary and depressed. Homosexuality was considered a mental illness, and those who acted on those impulses were committing illegal acts. Alan was arrested and charged with twelve counts of gross indecency under Section 11 of the Criminal Law Amendment Act of 1885. At the precinct, the unrepresented Turing was fingerprinted, photographed, and offered a choice: two years' imprisonment, as the law stated, or he could have his freedom, but would need to take estrogen therapy for a year, which was designed to reduce sexual urges. The hormonal shots would be paired with psychological counseling. Side effects included the growth of breasts, weight gain, and, as some would claim later, instability and depression.

Considered the father of modern computer science, Alan was a child prodigy. Born in Paddington, London, he grew up in various English homes, where neither he nor his older brother, John, was encouraged creatively. He had his first crush on a male classmate when he was in his teens. When the boy died suddenly from tuberculosis in 1930, Alan, stunned by the event, wrote the boy's mother for the next three years. He excelled in school, earning a distinguished degree in 1934, followed by a fellowship at King's College in 1935 and a Smith's Prize in 1936 for work on probability theory. A code breaker of sorts, he had a command of intellectual matters mixed with social awkwardness. His first major contribution was an idea for his Turing machine, his attempt to prove that machines can think. Acting like a human brain, the Turing machine would read a series of symbols from a tape, interpret them, and make "decisions" based on the findings. It was a very basic, yet unprecedented creation for an era before computers existed.

In 1938, he accepted a full-time job at the wartime cryptanalytic headquarters with a group of scientists hired to crack the Nazis' heretofore unbroken "Enigma" code. The team built a machine called the "bombes." At six and a half feet high and seven feet wide, and weighing a ton, it decoded the German message, and secured Alan a place in history.

Though painfully aware of his passion for men, he proposed to his colleague Joan Clarke in 1942, but quickly reneged after she accepted, admitting to her his homosexuality. His hunger for connection, love, and the touch of another man led him to admit to police in his living room the affair he'd had with a nineteen-year-old he'd met at a pub. Not one for discretion, the boy had told a friend, an unemployed felon, about the affair. The friend then stole things from Alan's home, assuming he wouldn't tell the cops for fear of being arrested.

Though Alan thought that opting for the estrogen shots would let him keep his job, his conviction led to a removal of his security clearance and prevented him from working for the government on cryptographic matters.

A year later he had another relationship, which led to a second run-in with the men in blue, or "the poor sweeties," as he used to call them. Already a known homosexual, he was considered a security risk. No longer working, he became more isolated in Manchester, retreating into himself and returning to his original passions: morphogenetic theory and the interpretation of quantum mechanics.

His life fell into a rhythm of consistency and routine. He continued working at Manchester University. He had friends over for dinner. He rode his bike to and from school. He ran marathons and reserved his weekly slot for use of the campus computers. Except on Tuesday, June 8, 1954, when he never showed at the computer center.

That morning, at 5:00 AM, his housekeeper arrived and saw from the window that his bedroom light was on and the lounge curtains open. Bottles of fresh milk were still on his apartment stoop, and the paper was stuck in the door. The housekeeper assumed he'd gone out early and forgotten to turn off the light. She picked up the milk and newspaper, let herself in as she'd done hundreds of times before, and, once inside, knocked on the bedroom door. Getting no answer, she entered. Alan's body was the first thing she saw, lying neatly on the bed. Froth covered his mouth. On the nightstand was a partially eaten apple. The housekeeper ran across the street to the neighbors' and phoned the police, who told her not to touch the body or anything in the home.

A classic sign of cyanide poisoning, the foaming mouth was an easy symptom for the pathologist to identify; the apple had been dipped in it. The pathologist clocked the time of death as late Monday night. The coroner confirmed the suicide as a deliberate act, blaming a disturbed mind. Jars of potassium cyanide were found in his apartment, along with a pair of recently purchased theater tickets and an invitation to a Royal Society function, which was happening the following night.

According to friends, Alan loved the Disney version of *Snow White* and could be spotted in Cambridge, weeks after the premiere, singing "Dip the apple in the brew / Let the sleeping death seep through." Many feel that the methodical man whose mind never shut off had been planning the suicide for months, but that his actually taking his life had been almost impulsive. Others say the genius concocted the perfect suicide. His greatest concern was protecting his mother. Fearing that his homosexuality had already caused her pain, and since suicide was still considered a crime, he hoped to hide the truth by making his death look like an accident. Since he was known for carelessly tooling around his university lab while crafting a variety of projects that involved potassium cyanide and gold-plating objects, she would often warn him about getting the deadly chemicals on his hands.

Alan's hunch was correct. His mother refused to accept the suicide verdict by the coroner, calling her son's death unintentional.

She cited the electrolytic experiment brewing in the back of her son's house; that weeks earlier he'd used the gold from his grandfather's watch to plate a teaspoon. She argued that he'd gotten cyanide on his hands and then put them in his mouth.

Unlike with Snow White, Alan's Prince Charming's kiss wouldn't happen until passage of the 1967 Act, which made homosexuality legal. Though it wasn't fully accepted and same-sex partners could not hold hands or show affection in public or in the presence of a third party, it was still a victory. He also missed the birth of the computer and the knowledge that every one of his grand ideas and predictions had come true.

UNEARTHED: One of the most common chemicals, not to mention fastest-acting poisons in the world, cyanide can be absorbed through the skin, ingested, or inhaled. Symptoms hit hard and fast. A lethal dose is estimated to be between fifty and three hundred milligrams. Within minutes one can experience rapid breathing, dizziness, weakness, headache, vomiting, and convulsions. As with carbon monoxide, cyanide can be colorless and odorless, though it's been described as having a bitter almond smell. Smoking cigarettes is probably one of the major sources of cyanide exposure. A lethal dose of cyanide is contained in 3.7 pounds of lima beans.

WARNING: Don't do mouth-to-mouth resuscitation just yet. If the suicide's lips or mouth still have cyanide or another poison on them, you could become the next victim.

DID YOU KNOW: Unavailable in the United States, paracetamol, an over-the-counter coal/tar-derived fever and pain reducer, is the most common drug used in overdose attempts, and accounts for almost seventy thousand cases a year in the United Kingdom.

THE APPLE: An iconic symbol, the apple has always been part of history. Aside from the nod to the fairytale "Snow White," and the Greek mythology that led Paris to fall in love with Helen and ignite a major war, the shiny fruit also started the mess between Adam and Eve and fell on Newton's head, and one of them a day will keep the doctor away. It's also the Apple corporation's logo. Many feel that Apple's logo, an apple with a bite taken out of it, is a secret nod to Alan, though the company denies this.

CAREER HIGHLIGHTS: Alan has gained recognition slowly over the past decades. Since 1966, the Association for Computing Machinery has given a yearly Turing Award. The year 1984 brought the TURING computer programming language, developed in Toronto. A road was named after him in Manchester. An official English Heritage blue plaque was placed on the Colonnade Hotel where he was born, and in 1994 a second was added to his home in Cheshire. A decade later, the University of Surrey placed a bronze sculpture of him on its campus, while another sculpture of him sitting on a bench with an apple in hand appears in the gay section of Sackville Park. A postage stamp issued by St. Vincent, West Indies, bears his image, and *Time* magazine highlighted the numbers specialist in their 1999 "Top 20 Scientists and Thinkers" issue. He also secured a spot in BBC's one hundred great British heroes. And in 2007, the Mathematics Department of Manchester University moved into its new Alan Turing Building.

Abbie Hoffman

BORN: November 30, 1936, Worcester, Massachusetts
DIED: April 12, 1989, New Hope, Pennsylvania
AGE: 52
METHOD: Overdose of barbiturates
DISCOVERED BY: His landlord
FUNERAL: More than one thousand people attended Abbie's memorial, which took place at his hometown of Worcester, Massachusetts. Among them were Chicago Seven alums David Dellinger and Jerry Rubin (who was also a Yippie cofounder), and Bill Walton, the radical Celtic of basketball renown. Afterward, mourners hit the streets in a symbolic singing peace march lead by Peter Seeger. A second tribute, called "No Regrets," took place at Manhattan's Palladium a few months later.
FINAL RESTING PLACE: Abbie was cremated, but where his ashes are scattered remains a mystery.

Personally, I always held my flower in a clenched fist. A semi-freak among the love children, I was determined to bring the hippie movement into a broader protest.
FROM HIS AUTOBIOGRAPHY, *SOON TO BE A MAJOR MOTION PICTURE*

Outspoken radical antiwar hippy activist Abbie Hoffman was many things: crude, harsh, reckless, and sexually promiscuous. Quiet and private he was not. As a public figure constantly creating havoc and vocalizing political injustices, he ironically chose an incredibly isolated and lonely way out.

Abbie sporting his notorious American flag shirt at an American flag–themed art show in New York, November 9, 1970

The 1960s was his decade. It was filled with war, racism, and unnecessary casualties. Abbie was going to change all that. In 1967 he co-formed the Yippies, the Youth International Party. He was also known for creating harmless political pranks. Most famous among them were crashing the New York Stock Exchange by throwing one-dollar bills onto the trading room floor from the viewing deck above, and trying to levitate the Pentagon, insisting that he and fifty thousand antiwar protesters could apply collective psychic power to change the thinking of Defense Department officials. A year later he became part of the Chicago Seven, which found him standing in front of a judge on charges of attempting to disrupt the Democratic National Convention with a riot. Over the span of twenty years, he was arrested forty times.

Abbie loved many things: parties, drugs, gambling. Though adored by all three of his lady loves, he often treated them poorly—verbally and physically abusing them. Yet, his minions craved his stunts and his hell-raising speeches, which often leaked out the conspiracies that Big Brother government was covering up.

He was born Abbott Howard Hoffman in Worcester, Massachusetts. In 1967 he created a pamphlet called *Fuck the System*, written under the pseudonym George Metesky. Caught between fact and fiction, his memoir *Revolution for the Hell of It*, which he wrote under the pseudonym "Free," was published in 1968. It recounted his radical escapades and became a guidebook for the day's "social and political activist." For most of the 1970s, he seemed lost. When he went underground in 1974, hoping to dodge an arrest for trying to sell $36,000 worth of cocaine, he also went under the knife, letting plastic surgery alter his features, specifically reshaping his nose. Though he disappeared as Abbie, he reemerged seven years later as Barry Freed, an environmental activist, and served a one-year prison term in a minimum-security facility.

The 1980s was not an easy decade for Abbie. The days of him strutting around in an American flag shirt had gone. The outspoken rebel, who was often called a comedian, had become a tragic figure. He was diagnosed with bipolar disorder, or manic depression, and in 1983 began having auditory hallucinations, which he said taunted him to kill himself. Listening to them, he took seventy-five Restorils he'd hoarded, wrote a suicide note, tore it up, scribbled another, and later woke up in Bellevue having had his stomach pumped. He had been earning a living giving speaking engagements, and though he insisted he'd take a year off, a week after being released from the hospital he was back on the circuit, making $60,000 a year delivering diatribes that had become meaningless and boring.

He dabbled for a bit in comedy, trying his hand at stand-up, but that didn't bring him the gigs or the money he hoped for. In 1986, he was arrested, with Amy Carter and thirteen others, for the forty-second time, while protesting CIA recruitment at the University of Massachusetts. The following year he fractured his leg in a car accident, which left him with a limp. His depression worsened. He switched drugs, trying a new wonder pill called Prozac, which seemed to make him even more depressed. The week before he died he stopped returning calls and seemed withdrawn.

On Monday, April 10, his younger brother Jack phoned, informing him that their mother had stage-four lymphoma, with only a month to live. When he ended the conversation telling Abbie not to leave him, that he was still considered the family patriarch, Abbie retorted, "Well, maybe it's time you took charge." Over the next two days, he cancelled a speaking job at Loyola College, stating that his mother was dying and he'd be unable to attend. He mailed two checks for tax payments, one to the IRS for $1,500 and one to his accountant for $4,000. Before sticking his favorite movie, *The Godfather,* into the VCR, he emptied 150 30-milligram phenobarbitals into a glass of Glenlivet single-malt Scotch whiskey, waited for the powder to dissipate, then drank the concoction. More shots of Scotch followed. He flushed the caplets down the toilet—earlier days might have found him doing the same with plastic baggies of coke before the feds showed up—and washed out the glass. Woozy and drunk, he headed into his bedroom, took off his shoes, and, still clothed, got into bed. He curled into a fetal position, placed his hands in prayer formation, resting them between his head and the pillow, and closed his eyes.

When Johanna, his longtime companion, and friends became worried, she called Abbie's landlord and asked him to check on him. Around 7:00 PM, he knocked on the door of the apartment (a converted turkey coop). When there was no answer, he peered in the window and saw him sleeping. He let himself in and found Abbie cold and lifeless. In his bedroom were more than two hundred pages of handwritten notes describing his plight with bipolar disorder. In the refrigerator was a box containing pot, hashish, trace amounts of LSD, and amphetamines.

Oddly, Abbie's death came at a time when a number of positive events were happening in his life. His last published article, on the October Surprise, had appeared six months earlier in *Playboy,* bringing the alleged Reagan administration conspiracy to the attention of a wide-ranging American readership for the first time, and rebooting his writing and activism career. Three days later he would have seen the two-year anniversary of his courtroom victory over the CIA. Eight months later he would have strutted down a red carpet for the film premiere of Oliver Stone's anti-Vietnam War film *Born on the Fourth of July*, in which he played a flag-waving radical who was gassed by state troopers on the steps of a college administration building.

He once told his ex-wife that he wanted to create "The Last Resort," an exotic place where people could "kill themselves in some grand style," adding that when he died, he wanted written on his tombstone, "He died in the streets." Neither of these things happened. Abbie's death was quiet and sad. He was tired, unable to think correctly, and devoid of passion. He outlived the 1960s, and the 1960s had long outgrown him.

UNEARTHED: Phenobarbital was brought to the market in 1912, by the drug company Bayer, which gave it the brand name Luminal. It remained a commonly prescribed sedative and hypnotic until the introduction of benzodiazepine in the 1910s, when, by accident, it was found helpful for use as tranquilizers for epilepsy patients. Between 1934 and 1945, German doctors under the Nazi Party used the drug to "euthanize" children born with disease or deformities. Their program's code name was Operation T-4.

Six years after Abbie's death, the British veterinary surgeon who became famous as the character Siegfried Farnon in the semi-autobiographical book and movie *All Creatures Great and Small* committed suicide at the age of eighty-four by injecting himself with an overdose of the drug.

CAREER HIGHLIGHTS: Aside from his constant misconduct and odd antics, Abbie Hoffman was a prolific writer. His works included *Revolution for the Hell of It; Woodstock Nation; Soon to Be a Major Motion Picture; Preserving Disorder: The Faking of the President; Vote! A Record, A Dialogue, A Manifesto—Miami Beach, 1972 and Beyond;* and *To America with Love: Letters From the Underground.* Most successful was his 1971 *Steal This Book.* A guide to living outside the establishment, it became a bestseller and earned a cult following. "It's embarrassing when you try to overthrow the government and you wind up on the bestseller's list," Abbie once said in an interview. *Steal This Book* was turned down by more than thirty publishers before Abbie collected $15,000 from friends to create Pirate Editions, so he could self-publish the book. It took months, but he finally found a distributor. Still, bookstores wouldn't carry it. Newspapers and TV and radio stations all refused to run advertisements for it. Still, *Steal* sold more than a quarter of a million copies between April and November 1971. Originally priced at $1.95, by the end of 1972, it was selling for $10.00 Today, an early edition will set you back $75 to $100. Although the New York Public Library has 9,993,000 books, it hasn't had an available copy of *Steal This Book* for twenty years. The Library of Congress doesn't have one, either.

Powerful People Unearthed

ANOTHER ON-AIR SUICIDE: Budd Dwyer, a politician and the state treasurer of Pennsylvania, in an almost copycat style to broadcaster Christine Chubbuck, pulled a similar stunt of public suicide in 1987 when, during a live televised press conference, he took out a gun and shot himself in the mouth. As the cameras rolled, his body fell against a wall, blood spilling out of his nose and mouth and flowing down the remains of his jaw and from the exit wound at the top of his head.

His name had been smeared when he was charged with receiving kickbacks of $300,000. Though offered a plea bargain, he refused, insisting he was innocent. Found guilty under state law, he was forced to resign from his position. He was facing fifty-five years in prison, along with a $300,000 fine. The day before his sentencing, Dwyer called a press conference, where many thought he'd be publicly stepping down from office. Instead, the show they witnessed was completely different. After reiterating his innocence, he read from a letter he had written.

> Since I'm a victim of political persecution, my prison would simply be an American gulag. I ask those that believe in me to continue to extend friendship and prayer to my family, to work untiringly for the creation of a true justice system here in the United States, and to press on with the efforts to vindicate me, so that my family and their future families are not tainted by this injustice that has been perpetrated on me.

As if for dramatic effect, he distributed envelopes to three of his staffers; one contained a suicide note for his wife, the second was an organ donor card, the third contained information relating to his case. The fourth envelope held a .357 magnum revolver. After asking those with weak stomachs to leave the room, he quickly stuck the gun in his mouth and pulled the trigger.

POLITICAL POWER: In 1933, House of Representatives member Samuel Austin Kendall, unhappy with the thought of leaving his job, shot himself while still in the House Office Building in Washington, D.C., before his successor was sworn in. In 1954, Getúlio Vargas, president of Brazil, shot himself during his impeachment trial in the presidential home. Others—José Manuel Balmaceda, president of Chile; Pierre Bérégovoy, French prime minister; and Vincent Foster, deputy White House counsel—all killed themselves with a firearm while still holding a political position. And after ordering his household to commit suicide in 1644, the last emperor of China's Ming dynasty, Chongzhen, hanged himself. Others—such as Nero, emperor of Rome; Otho, emperor of Rome; Maximian, emperor of Rome; Flavius Magnus Magnentius, emperor of Rome; Olaf I (Tryggvason), king of Norway; and Alexandros Korizis, prime minister of Greece—killed themselves after a military coup or defeat.

UNEARTHED: Hemlock, a poison also known as *Conium maculatum*, was used in ancient Greece to poison condemned prisoners. An overdose can produce muscular paralysis, loss of speech, and failure of the respiratory system, with death as an outcome. Known in southeastern Ireland as Devil's porridge, hemlock flourishes in the spring, when most other foliage is gone. It's also called poison parsley or spotted parsley, because of its resemblance to parsley. People refer to the red spots found on the stem and branches of the plant as "the blood of Socrates," who remains its most famous victim.

Eight

Fascinating Facts

Ten Questionable Deaths You've Heard of . . . and Ten You Might Not Have

Grace Jones once said "too much is never enough." For the following icons, accidental or not, too much clearly was.

It's often hard to determine if a death was planned or unintentional when one is on a slow and self-destructive path toward the end. Just as water in a pot over a flame will undoubtedly come to an angry, scalding boil, or a container of milk that starts off fresh will, past its expiration date, turn sour and become undrinkable, the nasty outcome is often no surprise. There isn't one person who could say he didn't see it coming. Not one friend who, when he heard about the death, was shocked.

For many in this list, suicide was not an impulsive or immediate act. Indeed, some made several suicide attempts before reaching permanent success. The dark murkiness makes it unclear what their true intentions were. Here, death is accompanied by conflict and unhappiness. The relationship with addiction, obliteration, and depression is deep, messy, and unattractive. With their decision-making process constantly clouded by drugs, and their bodies worn and weathered, who knows if that last shot of heroin, snort of coke, or swallow of vodka was enough this time to stop their hearts from pumping or hold hostage their last breath.

1. **Judy Garland** (1922–1969) was more than just a theatrical personality with a pair of ruby red shoes who longed to return home from Oz. She was one of the most beloved performers of stage and screen. At forty-seven years of age, she was found dead in her bathroom by her fifth husband, from an accidental overdose of barbiturates. She had attempted suicide before—once with a slight cut to her throat with a broken water glass. "All I could see ahead was more confusion. I wanted to black out the future as well as the past. I wanted to hurt myself and everyone who had hurt me," she said about that attempt. When her body was found in her rented London apartment, her blood contained the equivalent of ten 1.5-grain (97-milligram) Seconal capsules. More than 20,000 people stood outside the funeral home where her body lay, hoping to pay their respects.

2. **Elvis Presley** (1935–1977), with his twitching lip and southern rock-and-roll sensuality was found on the bathroom floor in his Graceland home by his fiancée. According to the medical investigator, the "King," while using the toilet, had fallen to the floor, then crawled several feet before dying. He had attempted

suicide in 1967; it wasn't unfathomable that he would try again. At the end of his life he was overweight and depressed. He took lots of pills; amphetamines, black beauties, and Dexedrine were among his favorites. More than one medical report indicated the cause of death as polypharmacy: too many different drugs in one's system—in Elvis's case, that would be fourteen. The toxicologist suggested cardiac arrhythmia. It was said Elvis spent $1 million a year on pills and doctors. His physician, George Nichopoulos, had prescribed more than ten thousand in the last eight months of the King's life. On the day of his funeral, more than a quarter of a million fans, celebrities, and members of the media waited in hopes of seeing an open casket exiting the grounds of Graceland.

3.   Mexican painter **Frida Kahlo** (1907–1954) lived her life in pain. After surviving a bout of polio as a child, she saw her health further ruined at eighteen when she was in a terrible bus accident, after which she had more than thirty operations. Though she supposedly died of a pulmonary embolism, no autopsy was performed, leaving many convinced that she'd committed suicide or died from an accidental overdose of drugs and liquor. A known alcoholic and drug user, she had attempted suicide before. Her last diary entry read, "I hope the exit is joyful—and I hope never to return." Her ashes are on display in her former home, La Casa Azul (The Blue House), in Coyoacán, Mexico.

4.   In 1971, The Doors front man **Jim Morrison** (1943–1971) moved to Paris, grew a beard, and made his last studio recording. He was found dead in his bathtub by his girlfriend, Pam Courson. Some

concluded that he died of heart failure brought on from high heroin intake. Many believe he committed suicide—addicted to drugs, overweight, and no longer with his band, he had much to be depressed about. In keeping with French law, an autopsy was never performed, since the medical examiner found no evidence of foul play. Jim's case was reopened when a new book written by a French journalist claimed that the singer had actually died from a heroin overdose in the bathroom of the Rock 'n' Roll Circus, a club on the Left Bank. The journalist asserted that Jim was carried home and put in his bathtub to deflect blame from the drug dealers he used. After Jim's death, Courson, also a fan of heroin, died of an overdose in 1974 in the Los Angeles apartment she shared with two male friends. Many say she started talking about seeing Jim soon. She was supposed to be buried next to him at Père Lachaise Cemetery in Paris, which was listed on her death certificate, but legal complications with transporting the body made that impossible. Instead, she was buried at Fairhaven Memorial Park, in California, under the name "Pamela Susan Morrison."

5. Wild child, heavy rocker, and famed singer **Janis Joplin** (1943–1970) let her fondness for heroin get the best of her when she accidentally OD'd in room 105 at the Landmark Hotel, in Hollywood, California. Janis had been staying there for the past six weeks, and on October 4 her body was found on the floor by her road manager. A cigarette was still held between her fingers. Her signature psychedelically painted Porsche was in the parking lot.

6. Uberinfluential singer-songwriter **Jimi Hendrix** (1942–1970) could noodle a guitar like nobody's business. Yet at the ripe old age of unlucky twenty-seven, like his two earlier mentioned musical friends, he was found dead in his London apartment by his girlfriend—a death that has never been fully clarified. Many feel he choked on his own vomit as a result of taking too many sleeping pills. Some sources say he washed down six German sleeping pills with wine, not knowing that the pills were ten times stronger than American ones. His girlfriend claimed the drugs were Vesperax and that he'd taken nine of them, but that he was alive when they entered their home. Yet police found Jimi alone—with no girlfriend in sight—dead for hours. In 1996, his girlfriend was found dead in her Mercedes-Benz at her cottage in England. As with Jimi, some say foul play; others insist on suicide.

7. Twenty-three-year-old actor **River Phoenix** (1970–1993) was a well-known star, and a closeted drug addict. In the wee hours on the morning of Halloween, after snorting some high-grade Persian Brown powder—a mix of crystal meth and opiates—in the bathroom of The Viper Room, a popular Los Angeles nightclub, he became violently ill. Taken outside by his brother, Joaquin, and actress Samantha Mathis, he collapsed and then convulsed on the sidewalk in front of the hot spot. The club became a temporary shrine as fans and mourners left flowers, pictures, and candles on the sidewalk and littered the walls with graffiti messages. Co-owned by actor Johnny Depp, the club remained closed for a week and, out of respect, is closed each year on Halloween.

8. Funnyman **John Belushi** (1949–1982), *Animal House* star and beloved original cast member of *SNL*, overdosed doing a speedball—a combination of cocaine and heroin, at the Chateau Marmont hotel (bungalow number 3) on Sunset Boulevard. Earlier that evening he had separate visits from Robert De Niro and Robin Williams.

9. Following in his hero Belushi's footsteps, *SNL*er **Chris Farley** (1964–1997) went on a four-day drug binge and spent his last hours with a hooker named Heidi, coked up and full of heroin. The end came when Chris collapsed in his Chicago apartment ten feet from the door as Heidi was exiting. His last words to her were, "Don't leave me." Not wanting to call for help, she took a photograph of him lying on the floor, then strutted out, closing the door behind her and leaving him to die. He was found the following day by his brother. Like Belushi, he was only thirty-three.

10. Australian Oscar-nominated actor **Heath Ledger** (1979–2008) was found naked and dead in the bedroom of his SoHo loft by his housekeeper and masseuse. Bottles of pills were on the nightstand and with him in the bed. Though many were quick to cite suicide, no note was found. As with Elvis, the toxicologist ruled the cause as acute intoxication by prescription drugs, including sleeping pills, pain relievers, and anti-anxiety medication.

Less illustrious, but still noteworthy, the following celebrities achieved pop culture status when they died—accidentally or intentionally.

1. Film actress, writer, and pin-up girl **Carole Landis** (1919–1948) suffered from depression and failed marriages—four of them. She attempted suicide in 1944 and 1946. With a fading career, she had an affair with fellow married actor Rex Harrison. The last person to see her alive, he discovered her dead body the following morning. He claims that he felt for a pulse, found one, and then fled the house. Police have a different story, saying that Carole had been dead for hours, having OD'd from Seconal. Two suicide notes were found, one for her mother, the other for Harrison. Later it was revealed that Harrison bribed a police officer to destroy the note. Carole's family insisted that Harrison had killed her and then made her death look like a suicide, but they were unable to prove this.

2. **Lizzie Siddal** (1829–1862) was a model—she was the model for Viola in Walter Deverell's *Twelfth Night*—and an artist from the pre-Raphaelite school who produced watercolors and drawings. Artist Dante Gabriel Rossetti used her as his muse and lover. By thirty she was unmarried, had given birth to a stillborn child, had lost Rossetti to another woman, and constantly fought chronic pain and flu-like symptoms. She was found by Rossetti in her bedroom, dead from an accidental overdose, or from suicide from laudanum, a liquid form of opiates. Rossetti, who felt responsible for her death, buried his poems in her coffin and made several attempts to speak with her via séances. A few years later he had her body exhumed so he could retrieve the poems he'd placed with it. Rumors say her body was perfectly preserved. Rossetti himself attempted suicide ten years after her death by using the same drug.

3. Film actor **Alan Ladd** (1913–1964), star of *This Gun for Hire* and *Shane*, was introduced to suicide when he watched his mother ingest ant poison. Years later, in 1962, a depressed alcoholic, he tried to do the same by shooting himself, but recovered from the wound. In 1964, he was found dead at his home in Palm Springs from a cerebral edema brought on by a mixture of alcohol and sedatives. His death was ruled accidental, though many feel it was a follow-up to his botched attempt two years earlier.

4. Suffering from work issues, depression, and insomnia, **Michael Reeves** (1944–1969), one of England's most promising young film directors, was found dead in his bedroom by his housekeeper, from a barbiturate overdose. His final film, *Witchfinder General*, is often considered one of Britain's greatest horror films.

5. **Edie Sedgwick** (1943–1971), the beautiful socialite and Andy Warhol favorite, was found dead by her husband in their bedroom. Though she had supposedly stopped drinking and taking drugs, her death certificate claims the immediate cause as "probable acute barbiturate intoxication" due to ethanol intoxication. Her alcohol level was registered at 0.17 percent, and her barbiturate level was 0.48 percent—sources say the latter of the two drugs was given to her by her husband. The coroner's report states undetermined/accident/suicide.

6. **Nick Drake** (1948–1974), England's darling singer-guitarist, suffered from drug dependency and had a nervous breakdown in 1972, after which he was hospitalized for five weeks. Depressed and dealing with insomnia, he had stopped washing his hair, cut-

ting his nails, and showering. In 1974 his mother found him lying across his bed, dead from an overdose of antidepressants. He'd left a suicide note for his girlfriend, but his family insisted the death was accidental.

7. Child actor **Scotty Beckett** (1929–1986) was best known for playing Spanky's friend in *Our Gang*, and Winky in *Rocky Jones, Space Ranger*. He lost his job on *Rocky* when jailed for a weapons charge, which was the start of his downward spiral. Failed marriages, violence, drugs, and alcohol, coupled with two car accidents, fueled his suicide attempts. Two days after checking into a nursing home for injuries caused from a beating, he was found dead in his room. A note and a bottle of sleeping pills were found near his body.

8. Blind Melon's **Richard Shannon Hoon** (1967–1995) left rehab with the understanding that a counselor from the program would accompany him on the band's tour. The goal of staying clean, sober, and out of jail was met for less than a month. After an all-night cocaine binge in New Orleans, Richard, who was sleeping in his bandmate's bed on the tour bus, was found by a roadie, who was unable to rouse the singer-songwriter. He had often joked that he'd surpassed the unlucky Twenty-seven Club of Janis, Jimi, and Jim, but in the end it was only by a few weeks.

9. Known as Kimberly Drummond on the popular 1980s sitcom *Diff'rent Strokes*, **Dana Plato** (1964–1999) had a long history with drugs and liquor, and a short acting career because of them. She

posed for *Playboy* and was arrested for armed robbery of a video store, claiming she needed money for rent. Her $13,000 bail was posted by Wayne Newton. She went to rehab and did some soft porn videos before being found dead in her mobile home outside her fiancée's parents' house in Oklahoma. The cause of death was ruled inconclusive: suicide or accidental OD on Carisoprodol (a muscle relaxant), Valium, and Lortab (a brand name for hydrocodone/acetaminophen).

10. By the age of twenty-five, **Brad Renfro** (1982–2008), best known for his role in *The Client*, had acted in twenty-one movies. An admitted heroin user in constant trouble with the law, he was found dead in his apartment early in the morning after a heavy night of partying. The Los Angeles County Coroner's Office stated that his death was an accidental overdose from booze, heroin, and morphine.

Ten Cases of Suicide . . . or Murder?

High-profile people paired with even higher-profile deaths makes for juicy material, rumors, investigations, and copious conspiracy theories. For these ten famed folks, many unanswered questions still surround their final moments. These cases remain unsolved and unresolved—thus we remain captivated. Many lament the lack of evidence or that the tests we have available today were unfortunately not yet invented at the time of these deaths. The best and the brightest—from forensic examiners to pathologists to blood pattern specialists to homicide detectives—still can't figure out when Marilyn Monroe's body was found; why Natalie Wood, with a known fear of drowning (though her husband later said she'd conquered her phobia), would have agreed to be on a yacht, let alone jump into the water, if she did in fact commit suicide; why George Reeves's hand was free of gunpowder if he pulled the trigger.

1. **Thelma Todd** (1905–1935), aka "the Ice-Cream Blonde," an actress who starred in more than one hundred films, including *Monkey Business* and *Horse Feathers*, was found in her car, which was idling inside her garage, dead from carbon monoxide poisoning. She'd donned a silk dress, a mink coat, and diamonds. An unidentified smudged handprint was on the door of her car and drops of blood were found inside, as well as on her mouth. The night she died she was attending a party thrown in her honor, where she fought with Roland West, her business partner and married lover. Even though the Los Angeles DA's office and a grand jury ruled the death a suicide, questions remained. Since no one thought of Thelma as depressed, many pointed fingers at

her husband. His abuse of the actress was well known, and the two were going through a nasty divorce. There were also rumors that mob boss Lucky Luciano wanted to use the rooms above Thelma's restaurant for gambling. When she told him, "Over my dead body," he'd replied, "That can be arranged." It was also said that Roland West, the last person to see Thelma alive, had purposely locked her out of her house, but closed the door to the garage without realizing she was inside, and she died as a result.

2. Actor **George "Superman" Reeves** (1914–1959) was found in his bed with a gunshot wound to his right ear. Earlier that evening, after having an argument at a restaurant with his fiancée, they returned home, where, against his "no guests after midnight" rule, they were met by friends at around 1:00 AM. Already in a bad mood, George went upstairs, but returned to the party later to complain about the noise. More drinking followed, and he went back upstairs with the intention of going to sleep. When the sound of a gunshot was heard by guests, one raced into his room to find him lying across his bed, naked, face up, his feet on the floor. A nine-millimeter Luger pistol was between his feet. Surprisingly, there were no gunpowder burns on his face and no fingerprints on the gun; gunshot residue testing was not commonly performed by the Los Angeles Police Department in 1959. Also, the bullet entered at an angle that would have made self-infliction improbable. Many asked, "Why suicide now?" George was preparing to marry his fiancée and travel to Spain and Australia, where he would have earned thousands of dollars from public appearances. He was scheduled to film more episodes of *Superman* and direct his own film. Some say his ex-lover, Toni

Mannix, might have been behind his death. She was enraged at his engagement and had made threatening phone calls to him. Her husband, MGM studio executive Eddie Mannix, might have arranged to have George killed. Some claim that he either put a bullet in the gun that usually contained blanks or managed to sneak into George's room, shoot him, and then escape.

3. Cherished actress and model, one-time wife of both Joe DiMaggio and Arthur Miller, and President Kennedy's favorite extracurricular activity, **Marilyn Monroe** (1926–1962) was more than an icon. She defined sexy in Hollywood. When she was found dead in her California home with a large amount of sedatives in her system, the autopsy showed no evidence of foul play. Marilyn had had four previous suicide attempts, along with a history of mood swings. The autopsy failed to show any sign of the sedatives in her stomach or intestinal tract, and there was no glass of water found at her bedside. Her death was classified as "probable suicide," but because of a lack of evidence, investigators could not call her death a suicide or homicide. One of the greatest unsolved cases, Marilyn's death is thought by some to be an accident due to a lack of communication between her pharmacist, her psychiatrist, and her internist—all of whom were prescribing her drugs. Others believe the FBI murdered Marilyn because she knew too much damaging information about the Kennedy brothers, having had an affair with one or both. The Mafia might have killed her as punishment to Robert Kennedy, who was cracking down on organized crime. Marilyn's death sparked one of the largest copycat effects. During August—the month she died—an additional 303 suicides in the United States took place, an increase of 12 percent.

4. Striking beauty and film actress **Natalie Wood** (1938–1981)—
   Rebel Without a Cause, West Side Story, Splendor in the Grass—had
   a paralyzing fear of drowning. On the night of her death, she
   was on her yacht with husband Robert Wagner and *Brainstorm*
   costar Christopher Walken, with whom Wagner suspected she
   was having an affair. The threesome had had dinner on the
   shore and had returned to the boat drunk. Natalie went down
   to her cabin while the two men stayed up to talk, and eventually an argument ensued. When her husband went to look for
   her, she was nowhere to be found. At about 1:00 AM, Wagner notified the harbor patrol, who recruited the Coast Guard's help
   to scout for her. At 7:30 AM, she was found floating facedown,
   dressed in a flannel nightgown and socks, some two hundred
   yards from the boat. The yacht's small dinghy was discovered
   adrift, with life vests aboard, some two hundred yards from her
   body. Though she had not taken her customary sleeping pills
   that evening, the autopsy showed traces of wine and pain medication in her system, along with bruises consistent with signs
   of a struggle. Her husband, who in his book *Pieces of My Heart:
   A Life* insists that her death was accidental, suspected that the
   dinghy was not docked correctly; he surmises that after hearing it bang into the side of the boat, Natalie had tried to secure
   it. Some believe she was upset with Wagner and the tension on
   the yacht that evening and was trying to go ashore. Others say
   she was disoriented and tipsy from wine and pills, and either
   accidentally or intentionally drowned. Fourteen years prior to
   her death, she was discovered by her secretary after attempting to overdose on pills, a fact that made many people assume
   this was another attempt.

5. **Paul Williams** (1939–1973), one of the founding members and original lead singer of The Temptations, was found in a deserted parking lot in Detroit, Michigan, on the ground near his car, dead from an apparent self-inflicted gunshot wound to the head. Though he suffered from depression and had personal and health problems, several pieces didn't add up. Paul had used his right hand to shoot himself in the left side of his head, and a bottle of alcohol was found near his left side, as if he'd dropped it while being shot. The gun had fired two shots, yet he was shot only once. Still, the coroner ruled his death a suicide.

6. When *Rebel Without a Cause* actor **Nick Adams** (1931–1968) didn't show for dinner, friend and lawyer Ervin Roeder went to his house, suspecting something was wrong with his usually prompt pal. He broke into Nick's Beverly Hills home to find him dead, propped up against the wall in his bedroom, dressed in jeans and a plaid shirt, inches from the phone. The autopsy showed he'd died from an overdose of the sedative Paraldehyde, which he took, in combination with a tranquilizer, to prevent tremors from his excessive alcohol consumption. The amount of the drug found in his system was enough to kill a person, but there were no traces of it found in his home. Despite his recent divorce and the failure of both his television shows, some believe that things were on an upswing for Nick, and that suicide wouldn't have been part of this plan. One friend attests he wouldn't have died without leaving a note for his children, for whom he had fought a long custody battle. Some suspect his lawyer Ervin Roeder of killing Nick over money. Others say the two men got into a fight, and Roeder, in order to calm Nick down, gave him the toxic con-

coction of pills and booze. When Nick OD'd, Roeder feared being blamed and told authorities that he'd found him like that.

7. Actor **Albert Dekker** (1904–1968) appeared in more than one hundred films, including *Dr. Cyclops* and *The Wild Bunch*. As a member of the California legislature, he spoke out against Red-baiting senator Joseph McCarthy, which got him blacklisted. Albert's son, who had killed himself at sixteen, was a loss he never got over. After not hearing from Albert for days, his fiancée went to his apartment to find notes from concerned friends taped to his door. After begging the landlord to let her in, she found Albert dead, kneeling in the bathtub, a hangman's noose around his neck. However, the knot was not tight enough for the rope to have strangled him, and he was wearing several belts and a set of handcuffs with the key attached. He was blindfolded with a scarf. Obscenities had been written on his body in lipstick; on his right cheek was the word *whip*, the phrase "Make me suck" was printed on his throat, and the image of a vagina had been drawn on his stomach. A hypodermic needle was sticking out of each arm. He had been dead for several days. Rather than investigate the death, the coroner ruled it accidental. While at first it seems Albert died from autoerotic asphyxiation, several elements left everyone unsettled: How had he managed to write so legibly on himself? If he was blindfolded, how could he have worked the chains around his body in such a manner? Where was the lipstick he'd used to write on himself? To deepen the mystery, $70,000 was missing from his house, as well as several electronic appliances.

8. American rhythm and blues singer-songwriter and pianist **Larry Williams** (1935–1980) was a chauffeur and a pimp before becoming well known for writing and recording rock-and-roll standards from 1957 to 1959. His hits included "Short Fat Fannie," "Bony Moronie," and "Dizzy Miss Lizzy." An arrest in the late 1950s for narcotics started his downward spiral, which was followed by more drugs, violence, and wild behavior. He drifted from record label to record label in the 1960s, recording a few songs, none of which became hits. In 1980, he was found dead in his Los Angeles home, from a gunshot wound to his head. The medical examiners claimed suicide, but others who knew Larry insisted that he was murdered because of his involvement in drugs, crime, and perhaps prostitution.

9. Two weeks after the Communist Party took over Czechoslovakia, Czech statesman **Jan Masaryk** (1886–1948) was found dead in the courtyard of the Foreign Ministry, having fallen from the bathroom window of his third-story apartment there. Though the initial investigation stated he'd committed suicide, it's now believed that he was pushed. A couch and radiator were in front of the window, plus investigators said that the window was difficult to open and only half the size of the one in his bedroom—making it an odd choice for a six-foot-tall, two-hundred-pound man. Forensic experts have questioned why there were feces on his body and the bathroom window, scratches on his hands and stomach, and signs of violence in the apartment; others who doubt that his death was a suicide mention that the loaded gun by his bed remained unused.

10. **Ray Johnson** (1927–1995), artist and friend of Andy Warhol, was dubbed "the most famous unknown artist in New York City" by the *New York Times* when his bloated body was fished out of Sag Harbor, on Long Island. Speculation was that he'd jumped from the Sag Harbor Bridge at a little past 7:00 PM, though no suicide note was found in the hotel he had checked into before he jumped, or in his house. He was discovered with $1,600 in his wallet and more than $400,000 in his bank account. When investigators took a closer look, a theme involving the number thirteen emerged, giving Ray a cult following. He died on Friday the 13th; he was 67 years old (6 + 7 = 13); and his hotel room was number 247 (2 + 4 + 7 = 13). Some, however, insist that jumping from the bridge had been a stunt, a live artistic piece, and that his death from it was merely accidental.

Not the Suite Life: Ten Hotel Suicides

More than free soap and an overpriced mini bar, a hotel room promises many things: anonymity, short-term ownership, and twenty-four-hour services. From the lowbrow motel room that merely gives shelter to the most extravagant suite that can furnish you with your own butler, people have always found hotels the perfect place to end their lives. With hotels, one's quest to live like a king "for he shall die tomorrow" is a dream not out of reach. A hotel's ultra high-end amenities can make one's last night on earth a pleasurable experience. Though having housed thousands before you, each day the room is stripped clean, ready for a new guest. In a hotel room you're alone, but not. There's solidarity in a building full of strangers. And strangers are the ones who find your body in a timely manner, preventing that task from falling to family and friends. A hotel also ensures privacy so that suicide attempts won't be interrupted. And let's not forget the hands-on clean-up crew. Housekeeping is sure to do a thorough job removing any evidence before the next guest can request an additional shampoo and conditioner.

According to A. K. Sandoval-Strausz in his book *Hotel: An American History*, suicides in hotels came into fashion as early as 1833, as hotels began to sprout up all across America. Self-killings were so common that the *New York Times* ran hotel-themed suicide stories yearly from 1850 through 1900.

In a *Time* magazine article published in 1943, suicides during wartime became so regular that one room clerk would ask registering guests, "Do you want a room to sleep in or to jump from?"

1. Pitcher **Win Mercer**, who enjoyed stints with the New York Giants and the Detroit Tigers, had just finished his barnstorming tour through the American West when he checked himself into San Francisco's Occidental Hotel under the name George Murray in 1903. According to the *New York Times*, the hotel's watchman was making his rounds and detected the odor of gas coming from Win's room. After breaking down the door, he found Win on the bed, his head covered with his coat. A tube ran from the gas jet and into his mouth; he was dead from carbon monoxide poisoning. His suicide note read, "Tell Mr. Van Horn of the Langham Hotel that Winnie Mercer has taken his life." Win had suffered from pulmonary problems; reports also blamed his suicide on gambling debts and intense relationships with women. His career record of 251 complete games ranks 77th in Major League history.

2. **F. O. Matthiessen**, a noted American literature critic and Harvard professor, leaped to his death from the twelfth story of the Manger Hotel in Boston on March 31, 1950. Since the death of his lover five years earlier, he'd become severely depressed. After a number of recurring dreams where he was jumping out a window and falling to his death, he turned his dream into a reality. On the hotel desk he left his Yale Skull and Bones pin, several letters he'd written to friends, a suicide note, and the keys to his apartment.

3. Fifty-four-year-old Dutch painter and pop star **Herman Brood**, an unruly alcoholic and junkie, committed suicide by jumping from the roof of the Amsterdam Hilton. A married man with

three children, he wrote in his suicide note only these words: "I don't feel like it anymore. Maybe I'll see you around." The 2007 film *Wild Romance* was based on his life.

4. Years after **Michael Hutchence**'s strange suicide in room 524 in Sydney's Ritz-Carlton, the hotel was renamed the Stamford Plaza. Today, the deluxe room with a view of the bay is one of the most requested suites and will cost you approximately $190 per night.

5. A favorite among drug addicts, alcoholics, and sexaholics is the famous Chelsea Hotel. **Sid** and **Nancy** both made room 100 famous when Sid found Nancy's bloody body in the room's bathroom in 1978, supposedly feet away from where he'd been sleeping. Room 100 still gets its share of requests, but the room was made part of a larger suite and now houses long-term guests.

6. In 1962, MGM costume designer **Irene Gibbons** committed suicide at the Knickerbocker Hotel in Los Angeles. Known for dressing famous actresses—Marlene Dietrich, Elizabeth Taylor, Claudette Colbert, Judy Garland, Lana Turner, and Doris Day—for their film roles, Irene checked into the hotel under an assumed name and slit her wrists. The wounds were not deep enough to do damage, so she leaped from the fourteenth-floor window. Her body landed on top of the hotel's awning.

7. The Skids's and Big Country's guitarist and singer-songwriter **Stuart Adamson** was found hanging in the closet of his hotel

room at Honolulu's Best Western Plaza Hotel in 2001. With his blood swimming with liquor, he tied a cord around his neck and wrapped the other end around the closet rail. He was found by a hotel housekeeper.

8. In 1979, **Donny Hathaway**, a singer, songwriter, and keyboardist most known for being Roberta Flack's duet partner of choice, checked into New York's Essex House and, after removing the glass from the window of his fifteenth-floor room, jumped. He was found dead on the sidewalk.

9. National Book Critics Circle Award–winning author **Michael Dorris** was found dead in 1997 at the Concord Motel in New Hampshire. Registered there under a false name, the author of *The Broken Cord* and *A Yellow Raft in Blue Water* swallowed a mixture of sleeping pills and vodka, and then tied a plastic bag over his head. Michael had been depressed for years. Ironically, that same day he was to have been honored at the twenty-fifth anniversary of the Native American Studies Program, which he had founded at Dartmouth. He was also to have been charged by the Hennepin County Attorney's Office with criminal sexual child abuse of some of his adopted children. His estranged wife, novelist Louise Erdrich, said he had attempted suicide multiple times, the first being on a Good Friday. Before finally taking his life, he left a note for housekeeping apologizing for the "mess."

10. Florida's Winter Park Quality Inn was host to The Band, who had just played a gig at the Cheek to Cheek Lounge. In 1986, around 2:30 AM, Canadian singer-instrumentalist **Richard Manuel**, who

had been hanging out with bandmate Levon Helm, suddenly said he needed to retrieve something from his room. Upon entering his quarters, he drained a bottle of Grand Marnier and then hanged himself while his wife slept only feet away.

DID YOU KNOW: A study performed by Dr. Paul Zarkowski, a psychiatrist in the Department of Psychiatry and Behavioral Sciences at the University of Washington, found that local residents who check into a local hotel room are twenty times more likely to kill themselves.

UNEARTHED: Something must be wrong with Chicago's Stevens Hotel, now called the Hilton, because of the high number of suicides that have occurred there. Mirroring the bad luck of Stephen King's Overlook Hotel, the Stevens saw five suicides occur there during a twelve-year period. Though Amelia Earhart, Babe Ruth, and Marshall Italo Balbo all stayed a night or two and lived to tell the tale, the following guests didn't: It all started in 1932, when Ethel Salhanick's crush didn't return her feelings, so she jumped out the window of her room at the Stevens and killed herself. Three years later, New Yorker Edwin Eder, who had suffered a nervous breakdown before traveling to Chicago, followed suit. Helen B. Martin, a Rogers Park housewife, checked into the hotel while her husband was away on business and took a leap of faith in 1937. She was later declared insane. A few months later, Flossie A. Castor, a twenty-seven-year-old grocery store bookkeeper, leaped to her death. And in 1947, suffering from mental illness, Florence Bear also jumped from her room window.

Ten Most Notable Suicides of Sports Figures

Sports figures have long been icons for us, thanks to their uber-endurance, speed, power, physical strength, and mental force. Whether springing into the air to slam-dunk a basketball, sending a baseball sailing into the sky past the outskirts of the stadium, performing ballet on ice, or pushing through a line of burly opposing teammates on the football field, athletes have moved from the courts, fields, stadiums, and into the category of famous figures.

1. Thrust into the spotlight when he was recruited to Scotland's national football team at age seventeen, **Hughie Gallacher** (1903–1957) met his end under the harsh criticism of the press. He became the national team's third most prolific scorer, although his football career was punctuated by indiscipline and controversial relationships. In 1957, he threw an ashtray at his son during an argument. Though his son was uninjured, Hughie was vilified by the press. The removal of his son by social services fed his depression, and on June 11, two trainspotters witnessed him crying and pacing before stepping into the path of an oncoming train.

2. Seven hours after he was found unresponsive in his suburban home, Pittsburgh Steelers guard **Terry Long** (1959–2005) was declared dead at forty-five. Though he was suffering from marital problems and financial bankruptcy, doctors blamed the swelling in his brain from head trauma sustained during his football career. However, a report filed three months later revealed that he had ingested antifreeze, which accounted for the swelling. A fan

of toxins, Terry had attempted suicide with rat poison in 1981, after being charged with violating the NFL's steroid policy.

3. At age twenty, **Michael Brent Adkisson** (1964–1987) was forced into the wrestling ring to replace his dead brother, David. A brawl with toxic shock syndrome in 1985 left him with brain damage and diminished physical strength. Often drunk and depressed, he left a suicide note for his family one April morning and drove to Lake Dallas, where he overdosed on alcohol and the tranquilizer Placidyl.

4. Lightweight boxer **Charles "Kid" McCoy** (1872–1940) was married ten times and had a burgeoning film career in Hollywood. But it was one of his whirlwind romances that would be his downfall. He was imprisoned for the murder of girlfriend Theresa Mors, and after being paroled in 1932, he was never the same. At the age of sixty-seven, he overdosed on sleeping pills. Part of his suicide note read, "Sorry I could not endure this world's madness."

5. Louisiana native **Richard Green** (1937–1983), who had been a Golden Gloves boxer himself, in the 1960s, was one of the first high-profile referee suicides. The incident that led to his deep depression was in not stopping a 1982 match between Ray Mancini and Duk Koo Kim, who in the fourteenth round, had clearly had enough. The twenty-three-year-old Duk died four days after the match, from injuries sustained in the ring. Unable to forgive himself, Richard shot himself one year later. Ironically, two other referees followed in his footsteps: Mitch Halpern shot himself

in 2000, and Toby Gibson died of carbon monoxide poisoning eight years later.

6. Born in Russia, NHL player **Roman Lyashenko** (1979–2003) played on the Dallas Stars and the New York Rangers, reaching the Stanley Cup Finals with the former. While on vacation in Turkey with his mother and sister in 2003, he was found dead in his hotel room. Unsuccessful at slitting his wrists, he'd hanged himself with a belt instead. He left a note apologizing for the suicide and claiming he had an incurable disease; the press at first surmised that this could be AIDS, but Turkish doctors ruled this out, along with any other severe illness.

7. More than twenty years after his last Major League Baseball game, **Arthur Irwin** (1858–1921), known and loved for reinventing the fielder's glove, disappeared on a steamer trip from New York City to Boston. A few days earlier, the Canadian American shortstop had learned he had an incurable disease. His death, which was ruled a suicide, ended up exposing the fact that he was illegally married to two women at the same time.

8. A tough rebounder and defender, Kansas City Kings basketball player **Bill Robinzine** (1953–1982) was found dead from carbon monoxide poisoning inside a rented storage unit by a caretaker of the Store-All storage lockers. Only twenty-nine years old, he was discovered slumped behind the wheel of his Oldsmobile Toronado, which sported the license plate "ROBY-1." A two-page suicide note was left for his wife explaining his deep sadness.

9. With nicknames such as the Rabid Wolverine and the Canadian Crippler and moves like the "dragon screw," professional wrestler **Chris Benoit** (1967–2007) enjoyed a prolific career with the WCW and WWF/E. All that ended over the span of a three-day weekend in June 2007, when he strangled his wife and suffocated his son. Using the cords from his weight machine, he then created a noose and attached over 550 pounds of weights to the other end to strangle himself. Autopsy studies of his brain revealed that several untreated concussions during his twenty-two-year wrestling career had left him with an advanced form of dementia.

10. **Peter Gregg** (1940–1980) first sold cars and then raced them. Nicknamed "Peter Perfect," he possessed an intense need for flawlessness in his racing, which took place during what was called the "golden age" of the Trans-Am Series. At age forty, he learned he had an incurable nervous system disorder that would disintegrate his physical abilities, and thus prevent him from driving. Without a desire to live, he shot himself in the head on a beach in Jacksonville, Florida. At the time of his death he had achieved a reputation as one of America's greatest and most successful road racers.

Ten of the Most Bizarre & Gruesome Suicides

With any collection of deaths, some examples stand out, whether because of an odd location, the strange way the person chose to take his life, the unusual objects he used, or the irony the world handed him at perhaps the worst time:

1. **Lou Tellegen** (1881–1934), a silent film actor who co-starred in several pictures and was romantically involved with Sarah Bernhardt, committed suicide by stabbing himself in the chest seven times with gold scissors engraved with his name. Bankrupt and depressed after a fire had ruined his face, he was found dead surrounded by film posters, photographs, and newspaper clippings.

2. **Florence Lawrence** (1886–1938) was a Canadian silent film actress who appeared in more than 270 films. Known to many as "the First Movie Star," "the Biograph Girl," or "the Girl of a Thousand Faces," she was badly burned in a studio fire while attempting to rescue someone in 1915. At twenty-nine years old, she took time off to recover, but never recaptured her star status. After three marriages that ended in either widowhood or divorce, and suffering from chronic pain from a rare bone marrow disease, she ingested ant paste, a poison. Though rushed to a hospital, she died a few hours later. Buried in an unmarked grave in the Hollywood Cemetery, she remained forgotten until 1991, when actor Roddy McDowall paid for her headstone, which was inscribed, "The Biograph Girl/The First Movie Star."

3. **Gig Young** (1913–1978), who appeared in almost sixty films

and won an Oscar for *They Shoot Horses, Don't They?*, was found dead by police in a Manhattan apartment next to the body of his fifth wife, a thirty-one-year-old German-born actress, to whom he had been married for only three weeks. In his hand was a gun—police assume he shot her first, then himself. It was later disclosed that Gig had been receiving psychiatric treatment—including experimental use of LSD from the controversial Dr. Eugene Landy, whose license was removed for his unorthodox methods of treating Beach Boy Brian Wilson. Ironically, Gig's last film was titled *Game of Death*.

4. **Nicolas Chamfort** (1741–1794), a French writer known for his witty epigrams and aphorisms, was usually in trouble with the law. Unable to face being imprisoned again, in 1793 he locked himself in his office and shot himself in the face, ripping off his nose and part of his jaw thanks to the pistol's malfunctioning. Panicked, he reached for a paper cutter and stabbed himself in the neck. That failed as well, and his artery remained uncut. As a last resort, he took the paper cutter and stabbed himself in the chest. In his own blood he wrote a note to the men who were going to arrest him.

5. Unpopular British politician **Robert Stewart** (1769–1822), who was bestowed with the title Viscount Castlereagh, suffered from fits of paranoia and instability, especially during the last year of his life. During an interview with English monarch King George IV, he admitted that a blackmailer was accusing him of paying men for sex. Three days later, he cut his throat with a letter opener and died. Recent scholarly speculation names syphilis as a possible reason for the viscount's sudden rages and rampant paranoia.

6. Buddhist monk **Thich Quang Duc** (1897–1963), in an act of protest against the persecution of South Vietnamese Buddhists, took a meditative sitting posture along a busy intersection on a Saigon road, covered himself in petrol, and lit himself on fire. Still in full meditative posture and showing no signs of suffering, he stayed perfectly still until he was burned beyond recognition. Pulitzer Prizes were given to journalists Malcolm Browne and David Halberstam for their coverage of the event. The body was re-cremated in an official ceremony, at which time it was discovered that Thich's heart remained perfectly intact.

7. American poet **Harold Hart Crane** (1899–1932), whose most notable works included *The Bridge* and *White Buildings*, was a volatile alcoholic. At noon, while heading back to New York from Mexico, he removed his topcoat and jumped off the steamship SS *Orizaba* and into the Gulf of Mexico, while shouting, "Goodbye, everybody!" The night before, he was badly beaten up for making sexual advances toward a male crew member. The lifeboats circled in vain for two hours before the steamship resumed its voyage. His body was never recovered. A marker on his father's tombstone in Garrettsville, Ohio, includes the inscription "Harold Hart Crane 1899–1932 LOST AT SEA."

8. **Lucius Annaeus Seneca** (4 BC–AD 65), a Roman philosopher and dramatist-poet, and adviser to the emperor Nero, was accidentally mixed up in the Pisonian conspiracy, a plot to kill Nero. When Nero found out, he insisted Seneca kill himself, allowing him to choose his method. Seneca opted for wrist slitting. Unfortunately, his blood was unusually thick and flowed slowly,

causing enormous pain. To speed up the process, he reached for poison, but that, too, failed. Without other options, he jumped into a pool of boiling water, suffocated on the steam, and drowned.

9. British actor and writer **Kenneth Halliwell** (1926–1967) was a mentor and partner to playwright Joe Orton. One evening, in their Noel Road flat in Islington, Kenneth took a hammer and smacked Joe nine times in the head. Afterward, he committed suicide by swallowing twenty-two Nembutals, sleeping pills. The bodies of both men were discovered the following morning, when a chauffeur came to collect Orton, who had a meeting with The Beatles regarding a screenplay he'd written for them. Kenneth's suicide note read, "If you read his [Orton's] diary, all will be explained. KH. PS: Especially the latter part"—a reference, it was believed, to Orton's cheating and promiscuity.

10. Comic writer and director **Clyde Bruckman** (1894–1955) wrote for films starring the Three Stooges and Abbott and Costello. His habit of recycling gags and material created for previous shows or comedians often got him into legal trouble, and as a result, producers were wary to hire him. Out of work and broke, he borrowed a forty-five-caliber pistol from Buster Keaton, claiming he needed it for a hunting trip. After eating a meal he couldn't afford at a café in Santa Monica, he went into the bathroom and shot himself in the head. On his body he left a typewritten note for his wife explaining that he'd chosen the location so he wouldn't dirty their home, and a letter to the police explaining that he couldn't pay for his funeral.

The Death of Cleopatra by Reginald Arthur, 1892

A Great Love: The Suicides of Cleopatra and Mark Antony

For some, death signifies a political statement or defines a particular period in history. For Cleopatra and Mark Antony, it was one of history's greatest acts of love.

Born in 69 BC, Cleopatra VII Thea Philopator was crowned Queen of Egypt at age seventeen and became a symbol of feminine power. In Rome, she lived openly as Caesar's mistress until his assassination. Aware of her unpopularity, she returned to Egypt, where she met Mark Antony, the Roman military activist who ruled Rome after Caesar's death, in the Second Triumvirate.

William Shakespeare, among many others, wrote about Cleopatra and Mark Antony, and numerous books, documentaries, and films have acknowledged their place in history—and led to the creation of another great love, that of Elizabeth Taylor and Richard Burton, who portrayed the prominent pair in a 1963 film.

The story goes like this: When Cleopatra's husband, Caesar, was killed, she quickly aligned herself with Mark Antony, becoming his mistress. He, along with Octavian and Marcus Lepidus, comprised the Triumvirate. Internal bickering led to a struggle for power, which in turn led to war.

Cleopatra had already retreated to her mausoleum and barricaded the door. Octavian's sister presented Antony with a fake suicide note she claimed was written by Cleopatra. After reading the letter, Mark Antony ordered his slave to kill him, feeling that life would be meaningless without her. Unable to grant his master's last wish, the slave killed himself instead, leaving Antony to fall on his own sword in a brokenhearted suicidal act. Though seriously injured, he didn't die. Cleopatra's slaves found him, placed him on a makeshift stretcher, and brought him back to her crypt. He was hoisted up through the window and supposedly died in her arms hours later. Inconsolable and out of her mind with grief, she decided to join him. She set the scene with the help of her servants, who were to die alongside her. Dressed in her finest royal garments, she lay down on a golden couch, regally placed a crown on top of her head, then took an asp (a poisonous snake), held it to her breast, and encouraged it to bite her. Another version says she bit herself and poured venom over the wound. A third story claims poisonous snakes were brought into her mausoleum in a water jar, while a fourth speaks of snakes hidden in flowers. Yet another suggests she scratched her arm with a poisonous hairpin. Regardless of the way the poison was taken, upon hearing the news, Octavian, the Roman emperor, sent for the Psylli, trained poison suckers, to revive her. Unfortunately, he was too late. Their tomb, said to contain the lovers buried side by side, has never been found.

DID YOU KNOW: Worshiped by the ancient Egyptians, the extremely poisonous Egyptian asp is a member of the cobra family. When threatened, the small, flat-headed snake—which can grow up to eight feet long—raises the front part of its body and spreads its neck into a hood, thus appearing more threatening to predators.

Grave Matters

It's been a long-accepted custom to pay homage to celebrities' headstones and visit their resting places. For decades, the resting places of Jimi Hendrix and Jim Morrison have ranked among the most visited, most beloved, and most decorated with mementoes and memorabilia. Celebrity suicides produce odd behavior and a protective instinct among fans and loved ones. Keith Richards admitted snorting his father's ashes, which he'd mixed with coke.

According to Dominic Maguire, spokesman for the National Association of Funeral Directors, many people feel unable to part with the remains of their deceased relatives. "Very often people keep the ashes of a loved one at home, possibly sitting on the mantelpiece in a nice ornate urn. It gives them a sense of nearness," he told *The Independent*, a UK newspaper and website, in 2000.

1. Post-punker **Ian Curtis**'s gravestone disappeared on July 1, 2008, from the Macclesfield cemetery in his hometown. Detectives said the curbstone, which had the inscription "Ian Curtis 18-5-80" and the words "Love Will Tear Us Apart," was taken sometime between Tuesday afternoon and Wednesday morning.

2. In June of 2008, Courtney Love claimed that **Kurt Cobain**'s ashes had vanished, along with a lock of his hair, from her secret hiding place in her closet. Having kept them in a pink teddy bear–shaped handbag, she said the thief must have been an acquaintance. The theft left Courtney feeling suicidal over her missing husband. To add to the weirdness, Natascha Stellmach,

an Australian artist who claimed she'd "acquired" Kurt's ashes, announced her intention to smoke the singer's remains. As part of an art installment, it was her goal to roll them into a spliff and puff—at an undisclosed location—at the close of her October 11 art exhibition in Berlin. The conclusion of this story remains mysterious: shortly after Stellmach made her claim, Courtney's rep said the ashes had not in fact been stolen. Regardless, some websites believe the artist went through with her plan, as many published articles on the subject after the eleventh either use the past tense to describe the event or don't say that Stellmach didn't succeed.

3. When Paula Yates was given **Michael Hutchence**'s ashes, she stitched them into a pink-and-white pillow, on which she slept. She supposedly kept the pillow on her bed while having sex with other lovers. A few years later, when she OD'd, she was cremated and her ashes were mixed with Michael's.

4. **Sylvia Plath**'s gravestone has been replaced a number of times because her fans keep scratching off the surname "Hughes," having blamed Ted Hughes for her downfall and death. Sylvia's full name now appears in bronze lettering on a replacement. Gravekeepers hope this will make it easier to repair.

5. Supposedly, while in the Heathrow terminal, **Sid Vicious**'s mother accidentally dropped his urn in the packed airport, sending most of his ashes up into the ventilation system for travelers to inhale. Others, however, say his ashes were scattered on Nancy Spungen's grave, in Philadelphia.

6. **Ernest Hemingway**'s tombstone needed to be repositioned and placed horizontally to prevent tourists from stealing the dirt as a souvenir.

7. Each year the Forest Hills Cemetery, a rural garden memorial park in Boston, offers an **Anne Sexton** tour and tribute. Taking place around the time of her death, the event offers readings by friends who knew her, followed by a discussion and a walk to her grave site.

Attempted Suicides

Dorothy Parker tried to kill herself four times. Übersexy French film actress and model Brigitte Bardot attempted suicide several times, the first time as a teenager, after her parents refused to permit her to marry director Roger Vadim until she was eighteen, and again at twenty-six, when she swallowed a bottle of sleeping pills and slit her wrists. "I took pills because I didn't want to throw myself off my balcony and know people would photograph me lying dead below," Bardot said.

Some who have attempted suicide have been outspoken, sharing the details of their plight in the hope of enlightening others. Some have talked about their momentary dance with death in a perfunctory manner, describing it as happening a long time ago, and claiming that they are in a much better place emotionally. Others have never offered the information freely, hoping to keep the tabloids from spilling their dirty little secret.

"An attempted suicide is not really an attempt at suicide in about 95 percent of cases," says George E. Murphy, M.D., a professor emeritus of psychiatry at Washington University School of Medicine in St. Louis. Murphy feels that attempters tend to use methods that allow for second thoughts or rescue. This confirms why many choose ineffective or slow-producing methods such as sleeping pills, a less invasive and less results-producing option versus hanging and firearms. This might also be why poison is the leading method of suicide attempts, followed by cutting at a close second.

Worldwide, an estimated ten to twenty million people attempt suicide yearly, with someone trying every thirty-four seconds. The following seventy-five celebrities, though highly successful in their careers, luckily failed when it mattered most.

1. **Maxene Andrews**: singer; 1954; pill overdose; devastated by the breakup of the vocal group she'd formed with her siblings, The Andrews Sisters.
2. **Adam Ant**: singer, actor; 1976; pill overdose; breakup with girlfriend.
3. **Mary Astor**: Oscar-winning actress; 1951; sleeping pills; battle with alcoholism. She maintained that it was an accident.
4. **Tai Babilonia**: Olympic skater; 1988; sleeping pills; failed skating performance at 1980 Olympics.
5. **Rona Barrett**: gossip columnist; 1971; sleeping pill overdose; depressed over the end of an eleven-year relationship, as well as her unending stream of work.
6. **Drew Barrymore**: actress; 1989; slashed wrists; stardom at a young age brought on nights of clubbing, alcohol and cocaine abuse, and two failed stints in rehab.
7. **Ludwig van Beethoven**: musician, composer; 1802; method unknown; increasing deafness had him desperate. He may also have suffered from bipolar disorder and lead poisoning.
8. **Fred "Rerun" Berry**: TV actor, *What's Happening!!*; before 1984 (three times); methods unknown; the dual financial boons and career pressures of fame saw him addicted to drugs and alcohol.
9. **Halle Berry**: Oscar-winning actress; 1996; carbon monoxide poisoning; devastated by the divorce from baseball player David Justice. What stopped her: the thought of her mother, who'd sacrificed so much for her, finding her.
10. **Danny Bonaduce**: child star; 2005; method unknown; during

the filming of his reality show *Breaking Bonaduce*, his wife asked for a divorce.

11. **Charles Bukowski:** German American poet, novelist; 1961; inhaling gas; motivation unknown.

12. **Maria Callas**: Greek opera singer; 1970; barbiturates overdose; unsuccessful in luring Aristotle Onassis away from then-wife Jackie Onassis. Later denied.

13. **Drew Carey**: actor, TV host; 1976, 1983 ("mid-twenties"); sleeping pill overdose; depression stemming from his father's death and molestation during his childhood.

14. **Martine Carol**: French actress; 1947; overdosed on alcohol and drugs, threw herself into the Seine; her affair with a married actor had ended catastrophically. The taxi driver who had driven her to the river saved her.

15. **Nell Carter**: TV star, *Gimme a Break!*; early 1980s (entered rehab in 1985); method unknown; was suffering from a cocaine addiction, and was devastated by her brother's bout with AIDS and the trauma of having been raped at the age of sixteen.

16. **Johnny Cash:** musician; 1967; exposure (in a cave near Chattanooga, Tennessee); addiction to alcohol, amphetamines, and barbiturates.

17. **Raymond Chandler:** crime novelist; 1955; pistol; Chandler was already suffering from alcoholism and depression, but the death of his wife, Cissy, pushed him to attempt suicide. He failed; the bullet caused damage to the bathroom; his botched attempt was derided in newspapers all over the country.

18. **Gary Coleman:** child star; 1993; sleeping pills (twice); said he hated his life and the circumstances that he had put himself in.

19. **Judy Collins:** musician, singer; 1952; method unknown; intense pressure following her public debut with the Denver Symphony.

20. **Nadia Comeneci:** Olympic gymnast; 1978; drank bleach; intense pressure, in part from her parents' divorce.

21. **Sammy Davis Jr.:** entertainer (singer, dancer); 1958; gun; the mob had forced him to end his relationship with actress Kim Novak, so instead he married black dancer Loray White, but he was distraught at not being able to be with his true love.

22. **Walt Disney:** producer, director, writer, cofounder of the Walt Disney Company; 1932; alcohol, sleeping pills; motivation unknown. Wife Lillian found him, called doctor, had his stomach pumped.

23. **Micky Dolenz:** singer, member of The Monkees; early 1970s; sitting in front of traffic; both marriage and band relationships were collapsing.

24. **Olympia Dukakis:** actress; early 1950s ("college years"); method unknown; depression and family issues.

25. **Patty Duke:** actress; 1965–1969 (sometime during marriage to Harry Falk); method unknown; suffered from extreme manic depressive episodes due to undiagnosed bipolar disorder.

26. **Eminem:** rapper, actor; 1996; Tylenol overdose; separated from wife, Kim Mathers.

27. **Marianne Faithfull:** singer, actress; 1970; heroin overdose; had broken up with The *Rolling Stone*s's Mick Jagger, also lost custody of son with first husband.

28. **F. Scott Fitzgerald:** writer; 1935–1937 (twice); first method unknown, ingesting poison on second attempt; on top of debts and

illness, his novel *Tender Is the Night* had failed, and both Ernest Hemingway and the *New York Post* jeered him.

29. **Zelda Fitzgerald**: wife of F. Scott Fitzgerald; 1924 (approx.), 1930; sleeping pills, throwing herself in front of a train; depression, hallucinations. Both times, her husband found her and saved her life.

30. **Peter Fonda**: actor; 1950; .22 pistol; after his mother's suicide. Claims it was an accident.

31. **Clark Gable**: Oscar-winning actor; 1942; high-powered motorcycle; distraught over the death of wife, Carole Lombard, who had been killed in a plane crash while selling war bonds. Enlisted in the army as a tribute to her.

32. **David Gahan**: lead singer of Depeche Mode; 1995; slashing wrists with razor blade; besides a failed marriage, he was no longer finding joy in his work.

33. **Stan Getz**: saxophonist; 1954; drug overdose; police were after him for a failed attempt to rob a Seattle pharmacy.

34. **Dwight "Doc" Gooden**: MLB player; 1994; shotgun; had been suspended for two seasons after testing positive for cocaine use. Wife stopped him.

35. **Cary Grant**: actor; 1934; poison tablets; separation from wife, Virginia Cherrill; found by his servant. According to one source, Grant denied it was a suicide attempt.

36. **Mariette Hartley**: actress; 1994; contemplated jumping from the twenty-fourth floor of the building she was staying at; her father had killed himself in 1963, her mother had attempted suicide shortly thereafter, and Hartley was suffering from undiagnosed bipolar disorder. Recognizing that suicide ran in her family, Hartley thought of her children and couldn't

bring herself to do to them what her father had done to her.

37. **Susan Hayward**: Oscar-winning actress; 1955; method unknown; following custody battle with ex-husband Jess Barker over their twin children.

38. **John Hinckley Jr.**: attempted to assassinate President Reagan to impress actress Jodie Foster; 1981 (twice, both in prison); painkiller overdose, hanging—tied an army field jacket to the bar of his window and wrapped the sleeve around his neck; assassination failure combined with seven months of interrogation.

39. **Mark W. Hofmann**: forger, counterfeiter, murderer; 1987, 1988, 1990 (in prison); drug overdoses; the first was after his wife, Dorie Olds, filed for divorce, the second two occurred after a hearing during which the Board of Pardons refused to grant him parole. His first attempted suicide lost him the use of his forging hand: while unconscious, he lay on top of his right arm for twelve hours, and as a result the muscles atrophied.

40. **Whitney Houston**: R&B singer; 2005; tried to throw herself out of hotel window; reports of her motivation vary.

41. **Billy Joel**: singer, musician; 1970; drinking furniture polish; the failure of his band, Attila.

42. **Elton John**: singer, musician; 1970; put his head in a gas oven; the then-closeted singer was about to marry Linda Woodrow. Discovered by writing partner Bernie Taupin; the incident inspired their song "Someone Saved My Life Tonight."

43. **Shelley Long**: TV, movie actress; 2004; painkiller overdose; going through divorce with husband of twenty-two years. Long has denied that the overdose was a suicide attempt, calling it "an accidental overdose."

44. **Greg Louganis**: Olympic diver; 1972; slashed wrists; knee injury

ruined his dream of competing in the Olympics as a gymnast.

45. **John McCain**: senator, Republican presidential nominee for 2008 election; 1967–1972; method unknown; as POW in Vietnam, suffered five years of physical and psychological torture.

46. **Mindy McCready**: country singer; 2005; vodka, Ambien; suffered physical abuse from boyfriend William McKnight, who had once nearly strangled her.

47. **Jeanette MacDonald**: singer, actress; 1939; pill overdose; discovered that lover Nelson Eddy had gotten married. Discovered by co-star W. S. Van Dyke, who killed himself in 1943.

48. **Claude Monet**: French Impressionist painter; 1868; threw himself into the river Seine; for financial reasons—his first son's birth occurred while Monet was suffering from threat of poverty and starvation.

49. **Sinead O'Connor**: Irish singer; 1993, 1999; prescription drug overdose, twenty Valium tablets; bipolar disorder, custody battle with ex-lover over infant daughter.

50. **Tatum O'Neal**: actress; 1970s ("adolescence," before car accident in 1978—twice); first method unknown, slashing wrists with razor blade on second attempt; depression, molestation. Regarding her razor blade attempt, father Ryan O'Neal told her, "You cut them the wrong way, Tatum."

51. **Eugene O'Neill**: playwright, Nobel Prize winner; 1912; alcohol; recent divorce, other unknown reasons.

52. **Yoko Ono**: singer, wife of John Lennon; 1962; method unknown; after divorcing her composer husband Toshi Ichiyanagi, she returned to Japan but found herself lonely and unable to work.

53. **Jack Osbourne**: son of Ozzy Osbourne; 2000; sliced his hands with a broken bottle, chugged a bottle of absinthe, swallowed

Soma, Xanax, Dilaudid; depression compounded by the discovery that his girlfriend was cheating on him.

54. **Ozzy Osbourne**: rocker; 1962; made a noose out of his mother's clothesline, strung it up, and jumped from a chair; his parents' fighting, his father's alcoholism. Caught by his alcoholic father, who then beat him.

55. **Lee Harvey Oswald**: assassinated John F. Kennedy; 1959; cut his wrist; while traveling in the Soviet Union, found out that his application for Soviet citizenship had been denied.

56. **Tyler Perry**: playwright, actor, director, producer; 1985 (approximate: "teenage years"); slit his wrists; a physically and emotionally abusive father. After the suicide attempt, Perry changed his first name from his father's to Tyler.

57. **Edgar Allan Poe**: poet; 1848; overdose of laudanum; specific reason unknown, but he suffered from depression, madness, alcoholism, and poverty.

58. **Dennis Price**: British actor; 1954; left gas on in the oven at his apartment; alcoholism. Discovered by his servant.

59. **Richard Pryor**: Golden Globe–winning actor; 1980; fire that occurred while free-basing cocaine; unknown.

60. **Paul Robeson**: singer; 1961; slashed wrists; according to his son's claim, Robeson had been drugged with LSD by a CIA agent as part of a program called "MK Ultra."

61. **J. K. Rowling**: author, Harry Potter series; 1993 (mid-twenties); method unknown; struggling as a poor single mother and suffering from depression after separation from her husband, Portuguese journalist Jorge Arantes.

62. **Sidney Sheldon**: writer, producer, director, actor; 1934; alcohol and pill overdoses; because of poverty and constant relocation

due to his father's trouble in finding a job, Sheldon felt like an outsider. His father found him and later told him, "Life is like a novel. It's filled with suspense."

63. **Nina Simone**: jazz and blues singer; late 1970s; method unknown; a man she thought was a sponsor for her musical comeback in London turned out to be a con man who robbed and beat her.

64. **Diana Spencer**: Princess of Wales; 1982; threw herself down stairs while pregnant; unhappy marriage, bulimia.

65. **Donna Summer**: singer; 1976; tried to jump from hotel window; depression. Discovered by housekeeper.

66. **Elizabeth Taylor**: Oscar-winning actress; 1962; Seconal overdose; said that she "needed to get away," also, her contract with MGM had ended a few years prior and she hadn't made any films.

67. **Lana Turner**: actress; 1951; slit her wrists; in her latest film, *Mr. Imperium*, her singing had been overdubbed; depressed by this, she viewed her career as "a hollow success." She was discovered by her business manager, Benton Cole. Cole and MGM covered up her attempt by telling the press that she had fallen through the shower door.

68. **Tina Turner**: singer; 1968; method unknown; turbulent marriage to singer Ike Turner, who physically and emotionally abused her.

69. **Kurt Vonnegut**: genre-spanning writer; 1984; alcohol and pill overdose; in part as a reflection of his mother's suicide in 1944, after which he says "[Suicide] has always been a temptation for me." Someone saved him, but his attitude toward the attempt was always casual.

70. **Jean Wallace**: actress; 1946, 1949; sleeping pills, stabbing; husband Franchot Tone's affair with Barbara Payton, then hers and Franchot's divorce.

71. **Mike Wallace**: journalist; 1986; pill overdose; suffering from depression following a lawsuit in which Vietnam general William C. Westmoreland had accused Wallace of libel due to a documentary CBS did on the general.

72. **Tuesday Weld**: Oscar-nominated actress; 1955; aspirin, sleeping pills, gin; had fallen in love with a homosexual and was rejected.

73. **Hank Williams Jr.**: country singer-songwriter, musician; 1973; method unknown; drug and alcohol abuse.

74. **Owen Wilson**: Oscar-nominated actor, writer; 2007; slashed wrists, pill overdose; depression.

75. **Mary Wollstonecraft**: British writer, philosopher, feminist, mother of Mary Shelley (author of *Frankenstein*); 1795 (twice); first method unknown, jumping off the Putney Bridge into the Thames on second attempt; had discovered lover Gilbert Imlay's infidelities. The first time she was stopped by Imlay, the second time by strangers.

Selected Sources

The following websites were invaluable tools in my research: Wikipedia, the Internet Movie Database, the Encyclopedia of Death and Dying, About.com, Answers.com, Suite101.com, AlternativeReel.com, TheSmokingGun.com, HollywoodUSA.co.uk, Biography.com, the People's Almanac (trivia-library. com), Fuller Up: The Dead Musician Directory (elvispelvis.com/suicide.htm), Michelle Acker (www.writing-word.com/mystery/suicide.shtml), and You-Tube's collection of archival footage. In addition, we drew from a number of other helpful online libraries, news sources, suicide help centers, online archives, and fan- and family-constructed online memorials.

Sources for the book in general

General info

Alden, Raymond Macdonald. *Critical Essays of the Early Nineteenth Century: With Introduction and Notes* (1921). Whitefish, Montana: Kessinger Publishing, LLC, 2008.

Alvarez, A. *The Savage God: A Study of Suicide*. New York: Random House, 1972.

Anderson, Scott. "The Urge to End It All." *New York Times*, July 6, 2008.

Cutter, Fred. *Art and the Wish to Die*. Chicago: Nelson-Hall Inc., 1983.

Etkind, Marc. . . . *Or Not to Be: A Collection of Suicide Notes*. New York: Riverhead Trade, 1997.

Frasier, David K. *Suicide in the Entertainment Industry: An Encyclopedia of 840 Twentieth-Century Cases*. Ann Arbor, Michigan: University of Michigan Press, 2002.

Largo, Michael. *Final Exits: The Illustrated Encyclopedia of How We Die*. New York: HarperCollins, 2006.

_____. *Genius and Heroin: The Illustrated Catalogue of Creativity, Obsession, and Reckless Abandon Through the Ages*. New York: HarperCollins, 2008.

Leenaars, Antoon A. *Suicide, Suicide Notes, and Other Personal Documents*. New York: Human Sciences Press, Inc., 1988.

Maris, Ronald W., Alan L. Berman, and Morton M. Silverman. *Comprehensive Textbook of Suicidology*. New York: The Guilford Press, 2000.

National Institute of Mental Health. "Suicide in the U.S.: Statistics and Prevention." NIMH, http://www.nimh.nih.gov/health/publications/suicide-in-the-us-statistics-and-prevention.shtml (accessed March 9, 2009).

Rogers, James R., Jamie L. Bromley, Christopher J. McNally, and David Lester. "Content Analysis of Suicide Notes as a Test of the Motivational Component of the Existential-Constructivist Model of Suicide." *Journal of Counseling & Development* 85 (Spring 2007): 182–88.

Seinfelt, Mark. *Final Drafts: Suicides of World-Famous Authors*. Amherst, N.H.: Prometheus Books, 1999.

Stone, Geo. *Suicide and Attempted Suicide: Methods and Consequences*. New York: Carroll & Graf Publishers, 1999.

Thomas, Chris. "First Suicide Note?" *British Medical Journal*, July 26, 1980.

Ussher, James, Larry Pierce, and Marion Pierce. *The Annals of the World*. Green Forest, Arkansas: Master Books, 2003.

Wittkower, Rudolf, Margot Wittkower, and Joseph Connors. *Born Under Saturn: The Character and Conduct of Artists: A Documented History from Antiquity to the French Revolution*. New York: Random House, 1963.

Suicide Prevention Centers/Associations

American Association of Suicidology. "Suicide Prevention, Intervention, Research, Education, Training," http://www.suicidology.org (accessed March 2, 2009).

American Foundation for Suicide Prevention. "AFSP: Home," http://www.afsp.org/ (accessed March 2, 2009).

American Psychiatric Association. *American Psychiatric Association Practice Guidelines for the Treatment of Psychiatric Disorders: Compendium 2006*. Arlington, Virginia: American Psychiatric Publishing, Inc., 2006.

Canadian Association for Suicide Prevention. "CASP," http://www.casp-acps.ca/ (accessed March 2, 2009).

Centers for Disease Control and Prevention. "Suicide Prevention," http://www.cdc.gov/ViolencePrevention/suicide/index.html (accessed March 9, 2009).

Centre for Suicide Prevention. "Centre for Suicide Prevention: Information, Training, Research," http://www.suicideinfo.ca/ (accessed March 2, 2009).

Education Development Center, Inc. "Suicide Prevention Resource Center." Suicide Prevention Resource Center (SPRC), http://www.sprc.org (accessed March 2, 2009).

International Association for Suicide Prevention. "Home–IASP," http://www.iasp.info/ (accessed March 2, 2009).

Kastenbaum, Robert, ed. *Macmillan Encyclopedia of Death and Dying: Suicide Influences and Factors*. Woodbridge, Connecticut: Macmillan Reference USA, 2003.

NOPCAS. "nopcas.org." National Organization for People of Color Against Suicide, http://www.nopcas.com/ (accessed March 2, 2009).

Suicide Prevention Action Network USA. "SPAN USA," http://www.spanusa.org/ (accessed March 2, 2009).

Authors
Virginia Woolf

Bell, Quentin. *Virginia Woolf: A Biography*. London: Hogarth Press, 1972.

Bloomsbury Group. "Monk's House," http://bloomsbury.denise-randle.co.uk/monks_house.htm (accessed February 23, 2009).

Bond, Alma Halbert. *Who Killed Virginia Woolf?: A Psychobiography*. New York: Human Sciences Press, 1989.

Caws, Mary Ann, and Nicola Luckhurst, eds. *The Reception of Virginia Woolf in Europe (Reception of British Authors in Europe)*. London: Continuum, 2002.

_____. *Virginia Woolf*. New York: Overlook Press, 2004.

Curtis, Vanessa. *Virginia Woolf's Women*. Madison, Wisconsin: University of Wisconsin Press, 2002.

Dally, Peter. *The Marriage of Heaven and Hell: Manic Depression and the Life of Virginia Woolf*. London: Robson Books, 1999.

Hill-Miller, Katherine C. *From the Lighthouse to Monk's House: A Guide to Virginia Woolf's Literary Landscapes*. London: Duckworth Publishers, 2003.

Ingram, Malcolm. "Virginia Woolf: Suicide." Malcolm Ingram's Homepage, http://www.malcolmingram.com/suicide.htm (accessed February 23, 2009).

Johnson, Manly. *Virginia Woolf*. New York: Ungar Publishing Co., 1973.

Panken, Shirley. *Virginia Woolf and the "Lust of Creation": A Psychoanalytic Exploration*. New York: State University of New York Press, 1987.

Terr, Lenore C. "Who's Afraid of Virginia Woolf? Clues to Early Sexual Abuse in Literature." *Psychoanalytic Study of the Child* 45 (1990): 533–46, http://

www.pep-web.org/document.php?id=psc.045.0533a.

Woolf, Leonard. *Downhill All the Way: An Autobiography of the Years 1919 to 1939*. New York: Harcourt Brace & World, 1967.

ERNEST HEMINGWAY

Baker, Carlos. *Ernest Hemingway: A Life Story*. New York: Charles Scribner's Sons, 1969.

Graves, Neil A. "Remembering Papa: One Hundred Years After His Birth, Ernest Hemingway's Proud and Painful Legacy Endures." *Cigar Aficionado Magazine*, July/August 1999, http://www.cigaraficionado.com/Cigar/CA_Archives/CA_Show_Article/0,2322,331,00.html (accessed February 23, 2009).

"Hemingway Dead of Shotgun Wound; Wife Says He Was Cleaning Weapon." *New York Times*, July 3, 1961, http://www.nytimes.com/books/99/07/04/specials/hemingway-obit.html?_r=1 (accessed February 23, 2009).

Hemingway, Gregory. *Papa: A Personal Memoir*. Boston: Houghton Mifflin Company, 1976.

Hemingway, Leicester. *My Brother, Ernest Hemingway*. New York: Pineapple P, Incorporated, 1996.

Hotchner, A. E. *Papa Hemingway: The Ecstasy and Sorrow*. New York: William Morrow and Company, Inc., 1983.

Kohlmeier, R. E., M.D.; C. A. McMahan, Ph.D.; and V. J. M. DiMaio, M.D. "Suicide by Firearms: A 15-Year Experience." *The American Journal of Forensic Medicine and Pathology* 22, no. 4 (2001): 337–40, http://www.amjforensicmedicine.com/pt/re/ajfmp/abstract.00000433-200112000-00001.htm;jsessionid=JjdM2YPh971vTG4pdL4kVmVBFQ2Nnq9nyQbMLqX6CkyvWvpp5NnD!-858031623!181195628!8091!-1.

Mellow, James R. *Hemingway: A Life Without Consequences*. Boston: Houghton Mifflin Company, 1993.

Meyers, Jeffrey. *Hemingway: A Biography*. New York: Harper & Row, 1985.

Reynolds, Michael S. *Hemingway: The Final Years*. Boston: W. W. Norton & Company, Inc., 1999.

United Press International. "Hemingway Out of the Jungle; Arm Hurt, He Says Luck Holds." *New York Times*, January 26, 1954, http://www.nytimes.com/books/99/07/04/specials/hemingway-jungle.html (accessed February 23, 2009).

HUNTER S. THOMPSON

Allen, Henry. "Last Words: A Testament to Hunter S. Thompson." *Washington Post*, September 9, 2005, C01, http://www.washingtonpost.com/wp-dyn/content/article/2005/09/08/AR2005090801993.html (accessed February 23, 2009).

Brinkley, Douglas. "Football Season Is Over." *Rolling Stone*, September 8, 2005.

Greenwood, Sterling. "Hunter's Friends Say Hollywood Hijacked His Funeral." *Aspen Free Press*, August 19, 2005, http://www.aspenfreepress.com/funeralhijack.html (accessed February 23, 2009).

Jordison, Sam. "Following Hemingway to the Grave." *The Guardian*, July 10, 2007, http://www.guardian.co.uk/books/booksblog/2007/jul/10/following hemingwaytothegra (accessed February 23, 2009).

KRT. "Last Journey Echoes Hemingway." *Sydney Morning Herald*, February 22, 2005, http://www.smh.com.au/news/Books/Last-journey-echoes-Hemingway/2005/02/22/1108834781484.html (accessed February 23, 2009).

Larocque, Emily. "Remembering Gonzo: How I Survived Hunter S. Thompson, Aspen's Best-loved and Most Notorious Resident." *Outside Magazine*, November 2006.

Lombardi, John. "Death by Gonzo." *New York*, July 21, 2008.

Millard, Rosie. "Hunter S. Thompson: Our Crazy Gonzo Life." *The Sunday Times,* March 23, 2008, http://women.timesonline.co.uk/tol/life_and_style/women/relationships/article3602025.ece (accessed February 23, 2009).

"Obituary: Hunter S. Thompson." *BBC News*, February 21, 2005, http://news.bbc.co.uk/2/hi/entertainment/4283349.stm (accessed February 23, 2009).

"Obituary: Hunter S. Thompson." *The Economist*, February 24, 2005, http://www.economist.com/obituary/displaystory.cfm?story_id=3690414 (accessed February 23, 2009).

Perry, Paul. *Fear and Loathing: The Strange and Terrible Saga of Hunter S. Thompson*. New York: Thunder's Mouth Press, 1992.

Seelye, Katharine. "Ashes-to-Fireworks Send-Off for an 'Outlaw' Writer." *New York Times*, August 22, 2005.

Wenner, Jann S., and Corey Seymour. *Gonzo: The Life of Hunter S. Thompson*. Boston: Little Brown & Co., 2007.

Whitmer, Peter O. *When the Going Gets Weird: The Twisted Life and Times of Hunter S, Thompson*. New York: Hyperion, 1993.

SYLVIA PLATH

Aird, Eileen M. *Sylvia Plath*. New York: Harper & Row, 1974.

Alexander, Paul. *Rough Magic: A Biography of Sylvia Plath*. New York: Viking Press, 1991.

Axelrod, Steven Gould, and Nan Dorsey. "The Drama of Creativity in Sylvia Plath's Early Poems." *Pacific Coast Philology* 32, no. 1 (1997): 76–86, http://www.jstor.org/stable/1316781.

Becker, Jillian. *Giving Up: The Last Days of Sylvia Plath*. New York: St. Martin's Press, 2003.

Butscher, Edward. *Sylvia Plath: Method and Madness*. New York: Seabury Press, 1976.

Hayman, Ronald. *The Death and Life of Sylvia Plath*. Charleston, S.C.: The History Press, 2003.

McManamy, John. "Sylvia Plath—In Her Own Words." McMan's Depression and Bipolar website, http://www.mcmanweb.com/sylvia_plath.html (accessed February 23, 2009).

Rosenblatt, Jon. "Sylvia Plath: The Drama of Initiation." *Twentieth Century Literature* 25, no. 1 (1979): 21–36, http://www.jstor.org/stable/441398.

Stevenson, Anne. *Bitter Fame. A Life of Sylvia Plath*. Boston: Mariner Books, 1998.

Wagner-Martin, Linda. *Sylvia Plath: A Biography*. New York: Simon & Schuster, 1987.

ANNE SEXTON

Fox, Margalit. "Diane Wood Middlebrook, Biographer, Dies at 68." *New York Times*, December 17, 2007.

Furst, Arthur. *Anne Sexton: The Last Summer*. New York: St. Martin's Press, 2000.

Harvard Square Library. "Anne Sexton," http://harvardsquarelibrary.org/poets/sexton.php (accessed March 9, 2009).

Marx, Patricia. "Interview with Anne Sexton." *The Hudson Review* 18, no. 4 (1965): 560–70, http://www.jstor.org/stable/3849705.

Middlebrook, Diane Wood. *Anne Sexton: A Biography*. Boston: Houghton Mifflin, 1991.

Morrow, Lance. "Pains of the Poet—and Miracles." *Time*, September 23, 1991.

Reich, Angela. "Review: Anne Sexton: Discipline Forced Upon Madness." *Contemporary Literature* 33, no. 3 (1992): 556–62, http://www.jstor.org/stable/1208483.

Salvio, Paula M. *Anne Sexton: Teacher of Weird Abundance* (SUNY Series, Feminist Theory in Education). New York: State University of New York Press, 2007.

Sexton, Anne. *Anne Sexton: A Self-Portrait in Letters*. Edited by Linda Gray Sexton and Lois Ames. Boston: Houghton Mifflin Harcourt, 2004.

Sexton, Linda Gray. *Searching for Mercy Street: My Journey Back to My Mother*. London: Little, Brown, & Co., 1995.

University of Texas at Arlington. "Anne Sexton: A Brief Biography," http://www.uta.edu/english/tim/poetry/as/bio1.html (accessed March 9, 2009).

ACTORS

SPALDING GRAY

Boland, John. "Spalding Gray: Home," http://www.spaldinggray.com/home.html (accessed February 16, 2009).

Casey, Nell. "For Spalding Gray, One Last Tale." *New York Times*, May 28, 2006.

Demastes, William. *Spalding Gray's America*. New York: Limelight Editions, 2008.

Gray, Spalding. *Gray's Anatomy*. New York: Vintage, 1994.

"Interviews—Spalding Gray." *PBS* November 1999, http://www.pbs.org/wnet/newyork/series/interview/gray.html (accessed February 16, 2009).

McKinley, Jesse. "Spalding Gray, 62, Actor and Monologuist, Is Confirmed Dead." *New York Times*, March 8, 2004.

Miller, Kate. "Gray Noise: Walking with the Talking Man in New York." *Io* magazine, date unknown, http://www.altx.com/io/gray1.html (accessed February 16, 2009).

Perez, Hugo. "Desperately Still Seeking Spalding." *New York Times*, March 21, 2004.

Smalec, Theresa. "Spalding Gray's Last Interview." *PAJ: A Journal of Performance Art* 88, no. 30 (2008): 1–14.

Snead, Robin. "What It Feels Like . . . To Find Spalding Gray's Body." *Esquire* July 2007, http://www.esquire.com/dont-miss/wifl/spaldinggray0807

(accessed February 16, 2009).

Williams, Alex. "Vanishing Act." *New York Magazine*, February 2, 2004.

DOROTHY DANDRIDGE

Bogle, Donald. *Dorothy Dandridge: A Biography*. New York: Berkley Trade, 1999.

Dandridge, Dorothy, and Earl Conrad. *Everything and Nothing: The Dorothy Dandridge Tragedy*. New York: HarperCollins, 2000.

"Dorothy Dandridge Died of Pill Dosage, Coroner Now Says." *New York Times*, November 18, 1965, 54.

"Dorothy Dandridge Found Dead at Her Apartment in Hollywood." *New York Times*, September 9, 1965, 41.

"44-Word Handwritten Will of Miss Dandridge Filed." *New York Times*, October 12, 1965, 58.

James, Caryn. "After Climb to Stardom, a Tumble, Then Death." *New York Times*, August 20, 1999.

Leavy, Walter. "Who Was the Real Dorothy Dandridge?" *EBONY*, August 1999.

Mills, Earl. *Dorothy Dandridge: An Intimate Biography*. Los Angeles: Holloway House Publishing, 1999.

Watkins, Mel. "Crashing the Gates." *New York Times*, July 13, 1997.

Weinraub, Bernard. "Hollywood's First Black Goddess and Casualty." *New York Times*, August 15, 1999.

PEG ENTWISTLE

"And Who Is Peg Entwistle?" *New York Times*, February 20, 1927, X4.

Anger, Kenneth. *Hollywood Babylon*. New York: Dell, 1981.

Associated Press. "Peg Entwistle Dies in Hollywood Leap." *New York Times*, September 20, 1932: 2.

Austin, John. *Hollywood's Unsolved Mysteries*. New York: Shapolsky, 1990.

"English Actress with Guild." *Oakland Tribune*, May 5, 1929,

"Girl Leaps to Death from Sign." *Los Angeles Times*, September 19, 1932, A1.

Jacobston, Laurie. *Hollywood Heartbreak: The Tragic and Mysterious Deaths of Hollywood's Most Remarkable Legends*. New York: Simon & Schuster, 1984.

"Suicide Laid to Film Jinx." *Los Angeles Times*, September 20, 1932, A1.

Zeruk, James. "The Hollywood Sign Girl," http://thehollywoodsigngirl.com/home.html (accessed February 16, 2009).

DAVID STRICKLAND

"Actor Found Dead." *People*, April 1999, http://www.people.com/people/article/0,,615562,00.html (accessed February 16, 2009).

Baldwin, Kristen. "Sudden Tragedy." *Entertainment Weekly*, April 1999, http://www.ew.com/ew/article/0,,272971,00.html (accessed February 16, 2009).

"David Strickland." *The E! True Hollywood Story*, E! Network, June 4, 2000.

Levitan, Corey. "Check Out Those Who Checked In: Situation Rooms." *Las Vegas Review-Journal* (October 2007), http://www.lvrj.com/living/10298257.html (accessed February 16, 2009).

Ravo, Nick. "David Strickland, 29, Actor; Had Role in Television Sitcom." *New York Times*, March 24, 1999.

Wagner, Annie. " 'Suddenly Susan' Actor Commits Suicide." *South Coast Today*, April 1999, http://archive.southcoasttoday.com/daily/03-99/03-24-99/a02ae016.htm (accessed February 16, 2009).

MUSICIANS

IAN CURTIS

Corbijn, Anton. *Control*. Independent release, 2007.

Curtis, Deborah. *Touching from a Distance: Ian Curtis and Joy Division*. London: Faber & Faber, 2007.

Curtis, Natalie. ' "Suddenly the reality hit me.' " *The Guardian*, September 22, 2007, http://www.guardian.co.uk/film/2007/sep/22/periodandhistorical (accessed March 2, 2009).

"Investigating Domestic Violence Strangulation: Lethality." Blue Sheepdog, November 15, 2007, http://www.bluesheepdog.com/2007/11/15/ investigating-domestic-violence-strangulation-lethality/ (accessed March 2, 2009).

Lester, Paul. " 'It felt like someone had ripped out my heart.' " *The Guardian*, August 31, 2007, http://www.guardian.co.uk/music/2007/aug/31/ popandrock.joydivision (accessed October 18, 2007).

Middles, Mick, and Lindsay Reade. *Torn Apart: The Life of Ian Curtis*. London: Omnibus Press, 2006.

Morley, Paul. *Joy Division: Piece by Piece*. Medford, N.J.: Plexus Publishing, 2008.

Savage, Jon. "Joy Division: From Safety to Where?" *Melody Maker*, June 14, 1980, http://new.music.yahoo.com/blogs/rocksbackpages/99/joy-division-from-safety-to-where (accessed March 2, 2009).

KURT COBAIN

Berman, Alan L., David A. Jobes, and Patrick O'Carroll. "The Aftermath of Kurt Cobain's Suicide." In *Suicide Prevention: A Holistic Approach*, edited by D. De Leo, Armin Schmidtke, and René F. W. Diekstra. New York: Springer, 1998, 139–44.

Clarke, Martin, and Paul Woods. *Kurt Cobain: The Cobain Dossier*. Medford, N.J.: Plexus Publishing, 2006.

Cobain, Kurt. *Kurt Cobain: Journals*. New York: Penguin Group, 2003.

Cross, Charles R. *Heavier Than Heaven: A Biography of Kurt Cobain*. New York: Hyperion Books, 2002.

Kahn-Egan, Seth. "Nailed to the Pentad: A Dramatistic Look at the Death of

Kurt Cobain." Paper presented at the National Communication Association Convention, Chicago, November 1997, http://www.cla.purdue.edu/dblakesley/burke/kahnegan.html (accessed March 2, 2009).

Kennedy, Dana, and Benjamin Svetky. "Reality Bites." *Entertainment Weekly*, April 22, 1994, http://www.ew.com/ew/article/0,,301941,00.html (accessed March 2, 2009).

Kunkel, Benjamin. "Stupid and Contagious." *New York Times*, May 6, 2007.

Savage, Jon. "Kurt Cobain: The Lost Interview." *Guitar World*, 1997.

Strauss, Neil. "The Downward Spiral: The Last Days of Nirvana's leader." *Rolling Stone*, June 2, 1994.

Thompson, Dave. *Never Fade Away: The Kurt Cobain Story*. New York: St. Martin's Press, 1994.

Wallace, Max, and Ian Halperin. *Love & Death: The Murder of Kurt Cobain*. New York: Atria, 2005.

SID VICIOUS AND NANCY SPUNGEN

Bruno, Anthony. "Punk-Rock Romeo and Juliet: Sid Vicious and Nancy Spungen." truTV.com, date unknown, http://www.trutv.com/library/crime/notorious_murders/celebrity/sid_vicious/index.html (accessed March 2, 2009).

Butt, Malcolm. *Sid Vicious: Rock 'n' Roll Star*. Medford, N.J.: Plexus Publishing, 2005.

Cox, Alex, and Abbe Wool. *Sid and Nancy: Love Kills*. London: Faber & Faber, 1986.

Kifner, John. "Sid Vicious, Punk-Rock Musician, Dies, Apparently of Drug Overdose; Bail Is Reinstated." *New York Times*, February 3, 1979, 24.

ELLIOTT SMITH

D'Angelo, Joe. "Cause of Elliott Smith's Death Still Unclear, Coroner Says; No

Illegal Drugs Found." MTV.com, December 31, 2003, http://www.mtv
.com/news/articles/1483982/20031231/smith_elliott.jhtml (accessed
March 2, 2009).

_____. "Elliott Smith's Girlfriend Insists No Involvement in Death." MTV.com,
January 9, 2004, http://www.mtv.com/news/articles/1484246/20040109/
smith_elliott.jhtml?headlines=true (accessed March 2, 2009).

D'Angelo, Joe, and Rodrigo Perez. "One of Us Is on the Moon." MTV.com, 2007,
http://www.mtv.com/bands/s/smith_elliott/news_feature_102903/ (accessed March 2, 2009).

De Wilde, Autumn. *Elliott Smith*. San Francisco: Chronicle Books, 2007.

Di Maio, Vincent J. M., and Susanna E. Dana. *Handbook of Forensic Pathology*.
UK: CRC Press, 2006.

Hanft, Steve. *Strange Parallel*. Dreamworks SKG: 1998.

"Insane after His Wife's Death." *New York Times*, April 13, 1885.

Levy, Ariel. "The Long, Slow Death of Elliott Smith." *Blender*, January/February
2004.

Nugent, Benjamin. *Elliott Smith and the Big Nothing*. New York: Da Capo Press,
Inc., 2005.

Pareles, Jon. "Elliott Smith, 34, Rock Songwriter and Singer." *New York Times*,
October 23, 2003.

Ramirez, Charlie. "Sweet Adeline: The Official Elliott Smith Site—by Fans,"
http://www.sweetadeline.net/ (accessed March 2, 2009).

Redfern, Mark, and Marcus Kagler. "Better Off Than Dead, Elliott Smith
Comes Clean." *Under the Radar*, 2003, http://www.undertheradarmag
.com/es.html (accessed March 2, 2009).

Rutty, Guy N. *Essentials of Autopsy Practice: Recent Advances, Topics and Develop-
ments*. New York: Springer, 2004.

Smith, R. J. "Elliott Smith's Uneasy Afterlife." *New York Times*, July 18, 2004.

MICHAEL HUTCHENCE

Agar, Gerry. *Paula, Michael, and Bob: Everything You Know Is Wrong*. London: Michael O'Mara Books, Ltd, 2005.

Campbell, Caren Weiner. "The Devils Inside." *Entertainment Weekly*, November 23, 2001, http://www.ew.com/ew/article/0,,253239,00.html (accessed March 2, 2009).

Hutchence, Kelland, Susie Hutchence, Mario Ferrari, Jacqui Ferrari, and Dennis Patterson. "Official Michael Hutchence Memorial Website," http://michaelhutchence.org/ (accessed March 2, 2009).

Hutchence, Tina, and Patricia Glassop. *Just a Man: The Real Michael Hutchence*. London: Pan Books, 2001.

INXS Publications and Anthony Bozza. *INXS: Story to Story: The Official Autobiography*. New York: Simon & Schuster, 2006.

_____. *The Michael Hutchence Story*. New York: Bantam Press, 2005.

Owen, David. *Hidden Evidence: 40 True Crimes and How Forensic Science Helped Solve Them*. Ontario: Firefly Books, 2000.

Sheleg, Sergey, and Edwin Ehrlich. *Autoerotic Asphyxiation: Forensic, Medical, and Social Aspects*. Tucson, Arizona: Wheatmark, 2006.

Smolowe, Jill. "Fast Life, Sudden Death." *People*, October 2, 2000, http://www.people.com/people/archive/article/0,,20132453,00.html (accessed March 2, 2009).

St. John, Ed. *Burn: The Life and Times of Michael Hutchence and INXS*. London: Transworld, 1998.

ARTISTS

VINCENT VAN GOGH

Art History Archive. "Vincent Van Gogh—Biography, Quotes & Paintings," http://www.arthistoryarchive.com/arthistory/expressionism/Vincent-Van-Gogh.html (accessed March 9, 2009).

Blumer, Dietrich. "The Illness of Vincent van Gogh." *The American Journal of Psychiatry* (April 2002), http://ajp.psychiatryonline.org/cgi/content/full/159/4/519 (accessed March 9, 2009).

Brooks, David. "Vincent van Gogh: Chronology." The Vincent van Gogh Gallery, http://www.vggallery.com/ (accessed March 9, 2009).

De La Faille, J. B. *The Works of Vincent Van Gogh: His Paintings and Drawings Catalogue Raisonné*. New York: Reynal & Company, 1970.

Fell, Derek. *Van Gogh's Women: His Love Affairs and Journey into Madness*. New York: Carroll & Graf Publishers, 2004.

Hanson, Lawrence, and Elisabeth Hanson. *Passionate Pilgrim: The Life of Vincent van Gogh*. New York: Random House, 1955.

Kimmelman, Michael. "The Evolution of a Master Who Dreamed on Paper." *New York Times*, October 14, 2005, http://travel.nytimes.com/2005/10/14/arts/design/14kimm.html (accessed March 9, 2009).

Nagera, Humberto. *Vincent van Gogh: A Psychological Study*. UK: George Allen & Unwin Ltd., 1967.

Van Gogh, Vincent. *The Complete Letters of Vincent Van Gogh*. New York: Bulfinch, 2000.

Wolf, Paul. "Creativity and Chronic Disease: Vincent van Gogh (1853–1890)." *Western Journal of Medicine* 175 (5) (November 2001), http://www.pubmedcentral.nih.gov/articlerender.fcgi?artid=1071623 (accessed March 9, 2009).

DIANE ARBUS

Arbus, Diane. *Revelations*. New York: Random House, 1980.

———, ed. *Untitled: Diane Arbus*. New York: Aperture, 1995.

Arbus, Doon. "Diane Arbus, Photographer." *MS*, October 1972.

Bosworth, Patricia. *Diane Arbus: A Biography*. New York: Alfred P. Knopf, Inc., 1984.

Butler, Judith. "Surface Tensions: Judith Butler on Diane Arbus." *ArtForum*, February 2004.

DeCarlo, Tessa. "Diane Arbus: Revelations Beyond Shock." *The Brooklyn Rail*, April 2005, http://www.brooklynrail.org/2005/04/art/diane-arbus-revelations-beyond-shock (accessed March 9, 2009).

Goldberg, Vicki. "The Discomfort of Strangers." *Vanity Fair*, November 2003.

Goldman, Judith. "Diane Arbus: The Gap Between Intention and Effect." *Art Journal* 34, no. 1 (1974): 30–35, http://www.jstor.org/stable/775864

International Center of *Photography, Encyclopedia of Photography*. New York: Random House, 1984.

Israel, Marvin. "Diane Arbus." *Infinity*, November 1972.

Katzenstein, Bill. "Diane Arbus Revisited." *Shutter Release*, January 2004.

Lubow, Arthur. "Arbus Reconsidered." *New York Times Magazine*, September 14, 2003.

Magid, Marion. "Diane Arbus in New Documents." *Arts*, April 1, 1967.

Oppenheimer, Daniel. "Diane Arbus." *The Valley Advocate*, date unknown.

Sicherman, Barbara, and Carol Hurd Green, eds. *Notable American Women: The Modern Period: A Biographical Dictionary*. Cambridge, UK: Belknap Press, 1983.

Thurman, Judith. "Exposure Time: Diane Arbus's Estate Opens Up Her Life and Work to New Scrutiny." *The New Yorker*, October 13, 2003.

MARK ROTHKO

Breslin, James E. B. *Mark Rothko: A Biography*. Chicago: University of Chicago Press, 1993.

Canaday, John. *Embattled Critic: Views on Modern Art*. New York: Farrar, Straus, and Cudahy, 1962.

Danto, Arthur C. "Rothko's Material Beauty." *The Nation*, Vol. 267, December 21, 1998, http://www.questia.com/PM.qst?a=o&d=5002304184 (accessed March 9, 2009).

Fischer, John. "Mark Rothko: Portrait of the Artist as an Angry Man." *Harper's Magazine*, July 1970.

Glueck, Grace. "Mark Rothko, Artist: A Suicide Here at 66." *New York Times*, February 26, 1970.

Horsley, Carter B. "Mark Rothko." *The City Review*, September 19, 2005, http://www.thecityreview.com/rothko.html (accessed March 9, 2009).

Kimmelman, Michael. "ART REVIEW, Rothko's Gloomy Elegance in Retrospect." *New York Times*, September 18, 1998.

Kosoi, Natalie. "Nothingness Made Visible: The Case of Rothko's Paintings." *Art Journal* 64 (2005), http://www.questia.com/PM.qst?a=o&d=5010861663 (accessed March 9, 2009).

May, Stephen. "Rothko: Emotion in the Abstract." *World and I*, Vol. 13, July 1998, http://www.questia.com/PM.qst?a=o&d=5002295548 (accessed March 9, 2009).

Radic, Randall. "An Outsider in Latvia, America & Art: Mark Rothko." *Literary Traveler*, January 31, 2008, http://www.literarytraveler.com/literary_articles/rothko_latvia.aspx (accessed March 9, 2009).

Rothko, Mark. *The Artist's Reality: Philosophies of Art*. Edited by Christopher Rothko. New Haven, Conn.: Yale University Press, 2006.

Rosenblum, Robert. *On Modern American Art*. New York: Henry N. Abrams, Inc., 1999.

Sandler, Irving. *The Triumph of American Painting: A History of Abstract Expressionism*. New York: Harper & Row, 1976.

Seldes, Lee. *The Legacy of Mark Rothko: An Exposé of the Greatest Art Scandal of Our Century*. London: Secker & Warburg, 1978.

Selz, Peter. *Mark Rothko*. New York: Doubleday, 1961.

Shattuck, Kathryn. "Rothko Kin Sue to Transfer His Remains." *New York Times*, April 8, 2008.

Smith, Roberta. "GALLERY VIEW; For Rothko, It Wasn't All Black Despair."
New York Times, March 6, 1994.

POWERFUL PEOPLE

HITLER

Bullock, Alan. *Hitler: A Study in Tyranny*. New York: Harper & Row, 1962.

Franchetti, Mark. "Hitler's Burnt Bones Tipped into Sewer." *The Sunday Times*
(UK), October 1999, http://www.leesaunders.co.uk/html/world_war_II/
ww2_events/hitlers_remains.php (accessed February 16, 2009).

Galante, Pierre, and Eugene Silianoff. *Voices from the Bunker*. New York: Penguin
Group, Inc., 1989.

Irving, David. David Irving's Index to Items on Adolf Hitler, 1997, http://www
.fpp.co.uk/Hitler/ (accessed February 16, 2009).

Knopp, Guido, and Angus McGeoch (translator). *Hitler's Women*. London: Taylor & Francis, Inc., 2003.

Petrova, Ada, and Peter Watson. *The Death of Hitler: The Full Story with New Evidence from Secret Russian Archives*. New York: W.W. Norton & Co., 1995.

Trevor-Roper, Hugh. *The Last Days of Hitler*. Chicago: University of Chicago
Press, 1992.

Toland, John. *Hitler*. London: Wordsworth Editions, Ltd., 1997.

"Uneven Romance." *Time*, June 1959, http://www.time.com/time/magazine/
article/0,9171,864655-1,00.html (accessed February 16, 2009).

SIGMUND FREUD

AROPA. "Sigmund Freud—Life and Work," 1999, http://www.freudfile.org/
(accessed February 16, 2009).

Blumenthal, Ralph. "Freud: Secret Documents Reveal Years of Strife." *New York
Times*, January 24, 1984.

Breitbart, William, and Barry Rosenfeld. "Physician-Assisted Suicide: The

Influence of Psycho-Social Issues." Cancer Control. March 1999, http://
www.medscape.com/viewarticle/417699 (accessed February 16, 2009).

Decker, Hannah S. *Freud, Dora, and Vienna 1900*. New York: Free Press, 1992.

"Freud and Death." Time, July 1972, http://www.time.com/time/magazine/
article/0,9171,877882,00.html (accessed February 16, 2009).

Gay, Peter. Freud: A Life for Our Time. Boston: W. W. Norton & Co., 1988.

McCrone, John. "Dichotomistic: An Introduction to Organic Logic and Holistic Causality." Dichotomistic, http://www.dichotomistic.com/mind_
readings_freud.html (accessed February 23, 2009).

Schur, Max. *Freud: Living and Dying*. New York: International Universities
Press, 1972.

Webster, Richard. *Why Freud Was Wrong: Sin, Science, and Psychoanalysis*. New
York: Basic Books, 1996.

ALAN TURING

Gray, Paul. "TIME 100: Alan Turing." Time, March 1999, http://www.time
.com/time/time100/scientist/profile/turing.html (accessed February 16,
2009).

Hodges, Andrew. *Alan Turing: The Enigma*. New York: Simon & Schuster, 1983.

_____. "Alan Turing—Home Page," 1995, http://www.turing.org.uk/turing/
(accessed February 16, 2009).

Leavitt, David. *The Man Who Knew Too Much: Alan Turing and the Invention of the
Computer*. Boston: W. W. Norton & Company, Incorporated, 2005.

Teuscher, Christof. *Alan Turing: Life and Legacy of a Great Thinker*. New York:
Springer-Verlag, 2004.

Turing, Sara. *Alan M. Turing*. Cambridge, UK: W. Heffer & Sons, Ltd., 1959.

ABBIE HOFFMAN

Buckley, Tom. "The Battle of Chicago from the Yippies' Side." *New York Times*

Magazine, September 15, 1968.

Hoffman, Abbie. *Soon to Be a Major Motion Picture*. New York: Putnam, 1980.

———. *Steal This Book*. Cutchogue, N.Y.: Amereon Limited, 1976.

Jezer, Marty. *Abbie Hoffman: American Rebel*. New Brunswick, N.J.: Rutgers University Press, 1993.

King, Wayne. "Abbie Hoffman Committed Suicide Using Barbiturates, Autopsy Shows." *New York Times*, April 19, 1989.

———. "Mourning, and Celebrating, a Radical." *New York Times*, April 20, 1989.

Raskin, Jonah. *For the Hell of It: The Lives and Times of Abbie Hoffman*. Berkeley: University of California Press, 1997.

Sloman, Larry. *Steal This Dream: Abbie Hoffman and the Countercultural Revolution in America*. New York: Doubleday, 1998.

ANCIENT CASES

SOCRATES AND CLEOPATRA

Flamarion, Edith. *Cleopatra: The Life and Death of a Pharaoh*. New York: Harry N. Abrams, Inc., 1997.

Frey, R. G. "Did Socrates Commit Suicide?" *Philosophy* 53, no. 203 (1978): 106–108, http://www.jstor.org/stable/3749734.

Grant, Michael. *Cleopatra: A Biography*. New York: Simon & Schuster, 1972.

Griffin, Miriam. "Ancient Attitudes to Suicide." *The Classical Review* 42, no. 1 (1992): 130–32, http://www.jstor.org/stable/711930.

Higonnet, Margaret. "Suicide: Representations of the Feminine in the Nineteenth Century." *Poetics Today* 6, no. 1–2 (The Female Body in Western Culture: Semiotic Perspectives) (1985): 103–18, http://www.jstor.org/stable/1772124.

Marra, Realino, and Marco Orru. "Social Images of Suicide." *The British Journal of Sociology* 42, no. 2 (1991): 273–88, http://www.jstor.org/

stable/590371.

Southern, Patricia. *Antony & Cleopatra*. Gloucestershire, UK: Tempus, 2008.

Whitehead, David. "Two Notes on Greek Suicide." *The Classical Quarterly* 43, no. 2 (1993): 501–502, http://www.jstor.org/stable/639191.

Photo Credits

Page 14, suicide note. Courtesy of the Egyptian Museum, Berlin.

Page 26, Monk's House. Photo by Ben Ellis.

Page 31, the River Ouse. Photo by Ben Ellis.

Page 34, Virginia Woolf. Library of Congress, LC-USZ62-111438.

Page 37, Ernest Hemingway. Courtesy of the John F. Kennedy Library/U.S. National Archives and Records Administration.

Page 41, Ernest Hemingway's headstone. Photo by Kevin Dawson Jones.

Page 44, Hunter S. Thompson. Photo by Paul Harris/Getty Images Entertainment.

Page 57, Sylvia Plath. Photo by Eric Stahlberg, copyright © Smith College. Sylvia Plath Collection, Mortimer Rare Book Room, Smith College.

Page 64, Sylvia Plath Hughes's headstone. Photo by Danny Garside.

Page 68, Anne Sexton. AP Photo/Bill Chaplis.

Page 75, Anne Sexton's scrapbook. Copyright © Anne Sexton. Reprinted by permission of SLL/Sterling Lord Literistic, Inc.

Page 84, Spalding Gray. AP Photo/Craig Houtz.

Page 93, Peg Entwistle. Courtesy of the Bruce Torrence Collection/Hollywood Photographs.com.

Page 98, Hollywoodland sign. Courtesy of Bruce Torrence Collection/Hollywood Photographs.com)

Page 102, Dorothy Dandridge. AP Photo/Harold Filan.

Page 109, Dorothy Dandridge's body. Courtesy of Donald Bogle.

Page 111, David Strickland. Photo by Jeff Kravitz/FilmMagic/Getty Images.

Page 126, Ian Curtis. Photo by Chris Mills/Redferns/Getty Images.

Page 132, Ian Curtis's headstone. Photo by Michel Enkiri.

Page 150, Elliott Smith. Photo by Jon Super/Redferns/Getty Images.

Page 159, Michael Hutchence's coffin. AP Photo/ Katrina Tepper.

Page 175, Vincent van Gogh Café. Hulton Archive/Getty Images.

Page 182, Van Gogh brothers' headstones. Photo by Michel Enkiri.

Page 186, Diane Arbus. Photo by Roz Kelly/Michael Ochs Archives/Getty Images.

Page 193, Diane Arbus's obituary. Courtesy of A.D. Coleman. Copyright © 1971 by A.D. Coleman. All rights reserved. Reprinted with permission of *The Village Voice*.

Page 196, Mark Rothko. AP Photo.

Page 210, Socrates. Library of Congress, LC-USZ61-1503.

Page 215, Adolf Hitler. Courtesy of the U.S. National Archives and Records Administration.

Page 221, *News Chronicle* with headline on Hitler's death. Photo courtesy of Popperfoto/Getty Images.

Page 226, Sigmund Freud with his daughter. Library of Congress, LC-USZ62-85889.

Page 232, letter from Sigmund Freud. Hulton Archive/Getty Images.

Page 235, Alan Turing. *Life Magazine*/Time & Life Pictures/Getty Images.

Page 244, Abbie Hoffman. Photo by Tyrone Dukes/ Hulton Archive/Getty Images.

Page 284, *The Death of Cleopatra* by Reginald Arthur, 1892.